EXCEL® MANUAL

BEVERLY DRETZKE
University of Minnesota

ELEMENTARY STATISTICS: PICTURING THE WORLD

FIFTH EDITION

Ron Larson
Pennsylvania State University

Betsy Farber
Bucks County Community College

Prentice Hall
is an imprint of

PEARSON

The author and publisher of this book have used their best efforts in preparing this book. These efforts include the development, research, and testing of the theories and programs to determine their effectiveness. The author and publisher make no warranty of any kind, expressed or implied, with regard to these programs or the documentation contained in this book. The author and publisher shall not be liable in any event for incidental or consequential damages in connection with, or arising out of, the furnishing, performance, or use of these programs.

Reproduced by Pearson Prentice Hall from electronic files supplied by the author.

ISBN-13: 978-0-321-69380-8
ISBN-10: 0-321-69380-9

1 2 3 4 5 6 BRR 15 14 13 12 11

Prentice Hall
is an imprint of

www.pearsonhighered.com

▶ Contents:

	Larson/Farber *Elementary Statistics* Page:	*The Excel Manual* Page:
Getting Started with Microsoft Excel		1
Chapter 1 Introduction to Statistics		
Technology		
Example *Generating a List of Random Numbers*	34	21
Exercise 1 *Generating a List of Random Numbers for the SEC*	35	23
Exercise 2 *Generating a List of Random Numbers for a Quality Control Department*	35	25
Exercise 3 *Generating Three Random Samples from a Population of Ten Digits*	35	28
Exercise 5 *Simulating Rolling a Six-Sided Die 60 Times*	35	31
Chapter 2 Descriptive Statistics		
Section 2.1 Frequency Distributions and Their Graphs		
Example 7 *Using Technology to Construct Histograms*	46	35
Exercise 31 *Constructing a Frequency Distribution and Frequency Histogram for the July Sales Data*	50	40
Section 2.2 More Graphs and Displays		
Example 4 *Constructing a Pie Chart*	56	47
Example 5 *Constructing a Pareto Chart*	57	51
Example 7 *Constructing a Time Series Chart*	59	54
Section 2.3 Measures of Central Tendency		
Example 6 *Comparing the Mean, Median, and Mode*	68	56
Section 2.4 Measures of Variation		
Example 5 *Using Technology to Find the Standard Deviation*	84	60
Section 2.5 Measures of Position		
Example 2 *Using Technology to Find Quartiles*	101	61
Example 4 *Drawing a Box-and-Whisker Plot*	103	64
Technology		
Exercises 1 and 2 *Finding the Sample Mean and the Sample Standard Deviation*	121	67
Exercises 3 and 4 *Constructing a Frequency Distribution and a Frequency Histogram*	121	68

	Larson/Farber **Elementary Statistics** Page:	*The Excel Manual* Page:
Chapter 3 Probability		
Section 3.2 Conditional Probability and the Multiplication Rule		
Exercise 40 *Simulating the "Birthday Problem"*	155	75
Section 3.4 Additional Topics in Probability and Counting		
Example 1 *Finding the Number of Permutations of n Objects*	168	80
Example 5 *Finding the Number of Combinations*	171	81
Technology		
Exercise 3 *Selecting Randomly a Number from 1 to 11*	187	83
Exercise 5 *Composing Mozart Variations with Dice*	187	87
Chapter 4 Discrete Probability Distributions		
Section 4.2 Binomial Distributions		
Example 4 *Finding a Binomial Probability Using Technology*	206	93
Example 7 *Graphing a Binomial Distribution*	209	94
Section 4.3 More Discrete Probability Distributions		
Example 2 *Using the Poisson Distribution*	219	99
Technology		
Exercise 1 *Creating a Poisson Distribution with $\mu = 4$ for X = 0 to 20*	233	100
Exercise 3 *Generating a List of 20 Random Numbers with a Poisson Distribution for $\mu = 4$*	233	105
Chapter 5 Normal Probability Distributions		
Section 5.1 Introduction to Normal Distributions and the Standard Normal Distribution		
Example 4 *Finding Area Under the Standard Normal Curve*	242	107
Example 5 *Finding Area Under the Standard Normal Curve*	242	108
Section 5.2 Normal Distributions: Finding Probabilities		
Example 3 *Using Technology to Find Normal Probabilities*	251	110
Section 5.3 Normal Distributions: Finding Values		
Example 2 *Finding a z-Score Given a Percentile*	258	111
Example 4 *Finding a Specific Data Value*	260	113
Technology		
Exercise 1 *Finding the Mean Age in the United States*	299	114
Exercise 2 *Finding the Mean of the Set of Sample Means*	299	117
Exercise 5 *Finding the Standard Deviation of Ages in the United States*	299	118
Exercise 6 *Finding the Standard Deviation of the Set of Sample Means*	299	121

	Larson/Farber *Elementary Statistics* Page:	*The Excel Manual* Page:
Chapter 6 Confidence Intervals		
Section 6.1 Confidence Intervals for the Mean (Large Samples		
Example 4 *Constructing a Confidence Interval Using Technology*	308	123
Example 5 *Constructing a Confidence Interval, σ Known*	309	125
Section 6.2 Confidence Intervals for the Mean (Small Samples)		
Exercise 29 *Constructing a Confidence Interval for Miles per Gallon*	325	126
Section 6.3 Confidence Intervals for Populations Proportions		
Example 2 *Constructing a Confidence Interval for p*	329	128
Section 6.4 Confidence Intervals for Variation and Standard Deviation		
Exercise 11 *Constructing a 99% Confidence Interval*	341	131
Technology		
Exercise 1 *Using Technology to Find a 95% Confidence Interval*	351	133
Exercise 2 *Finding a 95% Confidence Interval for p*	351	136
Exercise 5 *Simulating a Most Admired Poll*	351	137
Chapter 7 Hypothesis Testing with One Sample		
Section 7.2 Hypothesis Testing for the Mean (Large Samples)		
Example 2 *Finding a P-Value for a Left-Tailed Test*	372	141
Example 3 *Finding a P-Value for a Two-Tailed Test*	372	142
Example 7 *Finding a Critical Value for a Left-Tailed Test*	377	144
Example 8 *Finding a Critical Value for a Two-Tailed Test*	377	145
Exercise 33 *Testing That the Mean Time to Quit Smoking is 15 Years*	383	146
Section 7.3 Hypothesis Testing for the Mean (Small Samples)		
Example 1 *Finding Critical Values for t*	388	149
Exercise 30 *Testing the Claim That the Mean Number of Hours is 11.0*	396	150
Section 7.4 Hypothesis Testing for Proportions		
Example 1 *Hypothesis Test for a Proportion*	399	152
Exercise 16 *Testing the Claim That 30% of Households Own a Cat*	402	153
Section 7.5 Hypothesis Testing for Variance and Standard Deviation		
Exercise 28 *Testing the Claim That the Standard Deviation is No More Than $30*	412	155
Technology		
Exercise 1 *Testing the Claim That the Proportion is Equal to 0.53*	423	156

	Larson/Farber *Elementary Statistics* Page:	*The Excel Manual* Page:
Chapter 8 Hypothesis Testing with Two Samples		
Section 8.1 Testing the Difference Between Means (Large Independent Samples)		
Exercise 31 *Testing the Claim That Children Ages 6-17 Watched TV More in 1981 Than Today*	437	159
Section 8.2 Testing the Difference Between Means (Small Independent Samples)		
Exercise 19 *Testing the Difference in Tensile Strength*	448	162
Section 8.3 Testing the Difference Between Means (Dependent Samples)		
Exercise 11 *Testing the Difference in Weight Before and After an Exercise Program*	457	164
Exercise 12 *Testing the Difference in Batting Averages Before and After a Baseball Clinic*	457	165
Exercise 23 *Constructing a 90% Confidence Interval for μ_D, the Mean Increase in Hours of Sleep*	460	166
Section 8.4 Testing the Difference Between Proportions		
Example 1 *A Two-Sample z-Test for the Difference Between Proportions*	463	169
Technology		
Exercise 1 *Testing the Hypothesis That the Probability of a "Found Coin" Lying Heads Up Is 0.5*	477	171
Exercise 3 *Simulating "Tails Over Heads"*	477	173
Chapter 9 Correlation and Regression		
Section 9.1 Correlation		
Example 3 *Constructing a Scatter Plot Using Technology*	486	175
Example 5 *Using Technology to Find a Correlation Coefficient*	489	178
Exercise 23 *Scatter Plot and Correlation for Hours Studying and Test Scores*	497	180
Section 9.2 Linear Regression		
Example 2 *Using Technology to Find a Regression Equation*	503	185
Exercise 20 *Finding the Equation of a Regression Line for Hours Online and Test Scores*	506	187
Section 9.3 Measures of Regression and Prediction Intervals		
Example 2 *Finding the Standard Error of Estimate*	516	188
Section 9.4 Multiple Regression		
Example 1 *Finding a Multiple Regression Equation*	524	191
Exercise 5 *Finding a Multiple Regression Equation, the Standard Error of Estimate, and R^2*	528	192

	Larson/Farber *Elementary Statistics:* Page:	*The Excel Manual* Page:
Technology		
Exercise 1 *Constructing a Scatter Plot*	537	194
Exercise 3 *Find the Correlation Coefficient for Each Pair of Variables*	537	197
Exercise 4 *Finding the Equation of a Regression Line*	537	198
Exercise 6 *Finding Multiple Regression Equations*	537	199
Chapter 10 Chi-Square Tests and the F-Distribution		
Section 10.2 Independence		
Example 2 *Performing a Chi-Square Independence Test*	554	203
Section 10.3 Comparing Two Variances		
Exercise 21 *Performing a Two-Sample F-Test*	572	206
Section 10.4 Analysis of Variance		
Example 2 *Using Technology to Perform ANOVA Tests*	579	207
Exercise 5 *Testing the Claim That the Mean Toothpaste Costs Per Ounce Are Different*	581	209
Technology		
Exercise 3 *Determining Whether the Populations Have Equal Variances—Teacher Salaries*	595	210
Exercise 4 *Testing the Claim That Teachers from the Three States Have the Same Mean Salary*	595	211
Chapter 11 Nonparametric Tests		
Section 11.1 The Sign Test		
Example 1 *Using the Sign Test*	600	213
Exercise 20 *Performing a Paired-Sample Sign Test*	606	215
Section 11.2 The Wilcoxon Tests		
Example 2 *Performing a Wilcoxon Rank Sum Test*	613	217
Section 11.3 The Kruskal-Wallis Test		
Example 1 *Performing a Kruskal-Wallis Test*	621	219
Section 11.4 Rank Correlation		
Exercise 6 *Testing a Claim about the Correlation Between overall Score and Price*	628	221
Technology		
Exercise 2 *Performing a Sign Test*	647	222
Exercise 3 *Performing a Wilcoxon Rank Sum Test*	647	223
Exercise 4 *Performing a Kruskal-Wallis Test*	647	225
Exercise 5 *Performing a One-Way ANOVA*	647	227

Getting Started with Microsoft Excel 2007

Overview

This manual is intended as a companion to Larson and Farber's *Elementary Statistics, 5th ed.* It presents instructions on how to use Microsoft Excel 2007 to carry out selected examples and exercises from *Elementary Statistics, 5th ed.*

The first section of the manual contains an introduction to Microsoft Excel 2007 and how to perform basic operations such as entering data, using formulas, saving worksheets, retrieving worksheets, and printing. All the screens pictured in this manual were obtained using the Office 2007 version of Microsoft Excel on a PC. You may notice slight differences if you are using a different version or a different computer.

Getting Started with the User Interface

GS 1.1	The Mouse

The mouse is a pointer device that allows you to move around the Excel worksheet and to select specific locations and objects. There are four main mouse operations: Select, click, double-click, and right-click.

1. To **select** generally means to move the mouse pointer so that the white arrow is pointing at or is positioned directly over an object. You will often select commands in the Office icon menu. Some of the more familiar of these commands are Open, Save, and Print.

2. To **click** means to press down on the left button of the mouse. You will frequently select cells of the worksheet and commands by "clicking" the left button.

3. To **double-click** means to press the left mouse button twice in rapid succession.

4. To **right-click** means to press down on the right button of the mouse. A right-click is often used to display special shortcut menus.

1

GS 1.2	The Excel 2007 Window

The figure shown below presents the top left side of the Exel 2007 window. The Office icon is located in the upper-left corner. Near the top of the screen, you see a row of several tabs: Home, Insert, Page Layout, Formulas, Data, Review, View, and Add-Ins. Each tab leads to a ribbon. The Home ribbon, shown below, presents groups of related commands. The groups that are displayed in the figure are Clipboard, Font, Alignment, and Number.

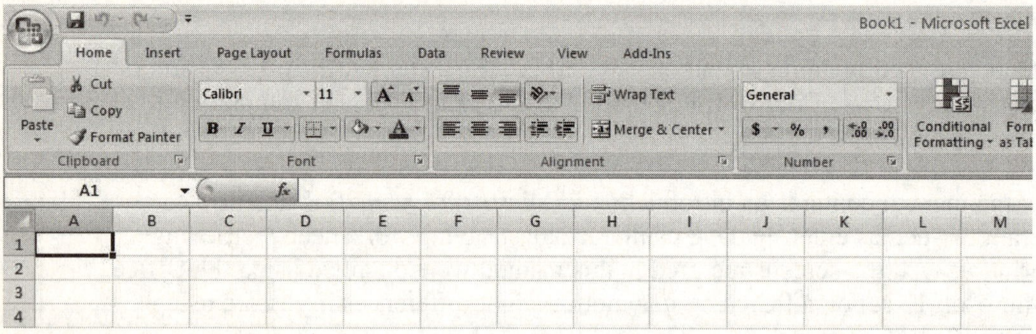

Clipboard commands may be replaced by keyboard shortcuts for those of you who prefer to use them. Some of the more familiar of these shortcuts are Ctrl+X for cut, Ctrl+C for copy, and Ctrl+V for paste.

GS 1.3	Ribbons and the Office Icon

When you click one of the tabs at the top of the screen, you will see a ribbon of related commands. The **Home ribbon** is displayed above in GS 1.2.

The **Insert** ribbon, shown below, includes groups of commands for Tables, Illustrations, Charts, Links, and Text.

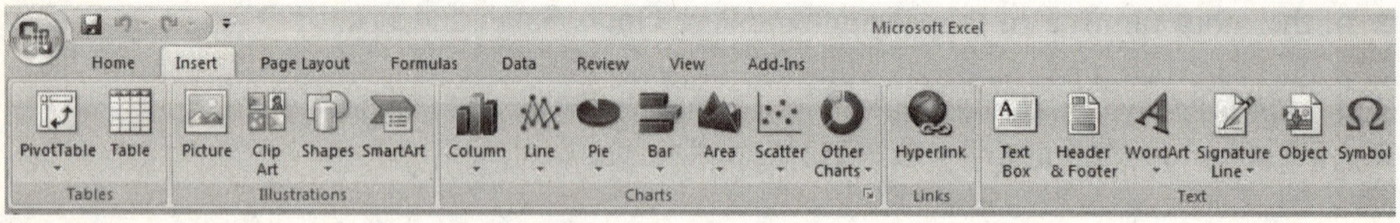

The **Page Layout ribbon** includes groups of commands for Themes, Page Setup, Scale to Fit, Sheet Options, and Arrange.

The **Formulas ribbon** includes groups of commands for Function Library, Defined Names, Formula Auditing, and Calculation.

The **Data ribbon** includes groups of commands for Get External Data, Connections, Sort & Filter, Data Tools, Outline, and Analysis.

The **Office icon** in the top left corner of the screen contains the commands New, Open, Save, Save As, Print, etc. Click the Office icon to display the menu.

Office icon

GS 1.4	Dialog Boxes

Many of the statistical analysis procedures that are presented in this manual are associated with commands that are followed by dialog boxes. Dialog boxes usually require that you select from alternatives that are presented or that you enter your choices.

For example, if you click the **Formulas** tab and select **Insert Function**, a dialog box like the one shown below will appear. You make your function category and function name selections by clicking on them. The SLOPE function that has been selected in this example is found in the Statistical category. If you want to find out what functions are available, you can type in a brief description of what you would like to do and then click the Go button.

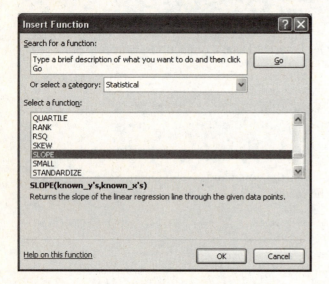

When you click the OK button at the bottom of the dialog box, another dialog box will often be displayed that asks you to provide information regarding location of the data in the Excel worksheet.

Getting Started with Opening Files

GS 2.1	Opening a New Workbook

When you start Excel, the screen will open to **Sheet 1** of **Book 1**. Sheet names appear on tabs at the bottom of the screen. The name "Book 1" will appear at the top.

If you are already working in Excel and have finished the analyses for one problem and would like to open a new book for another problem, follow these steps: First, at the top of the screen, click the **Office icon** and select **New**. Next, double-click **Blank Workbook** (or click **Create** in the bottom right corner). If you were previously working in Book1, the new worksheet will be given the default name Book 2.

The names of books opened during an Excel work session will be displayed in the Window group of the View ribbon. To return to one of these books, click the **View** tab at the top of the screen and select **Switch Windows** in the Window group. Then click the book name.

| GS 2.2 | Opening a File That Has Already Been Created |

To open a file that you or someone else has already created, click on the **Office icon** and select **Open**. A list of file locations will appear. Select the location by clicking on it. Many of the data files that are presented in your statistics textbook are available on the CD that accompanies this manual. To open any of these files, you will select the CD drive on your computer.

After you click on the CD drive, a list of folders and files available on the CD will appear. You will need to select the folder or file you want by clicking on it. If you have selected a folder, another screen will appear with a list of files contained in the folder. Click on the name of the file that you would like to open.

Getting Started with Entering Information

| GS 3.1 | Cell Addresses |

Columns of the worksheet are identified by letters of the alphabet and rows are identified by numbers. The cell address A1 refers to the cell located in column A row 1. The dark outline around a cell means that it is "active" and is ready to receive information. In the figure shown below, cell C1 is ready to receive information. You can also see C1 in the **Name Box** to the left of the **Formula Bar**. You can move to different cells of the worksheet by using the mouse pointer and clicking on a cell. You can also press [**Tab**] to move to the right or left, or you can use the arrow keys on the keyboard.

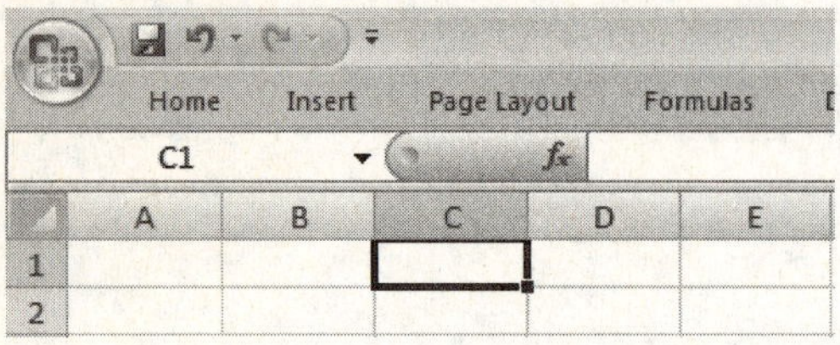

You can also activate a **range** of cells. To activate a range of cells, first click in the top cell and drag down and across (or click in the bottom cell and drag up and across). The range of cells highlighted in the figure below is designated B2:D4.

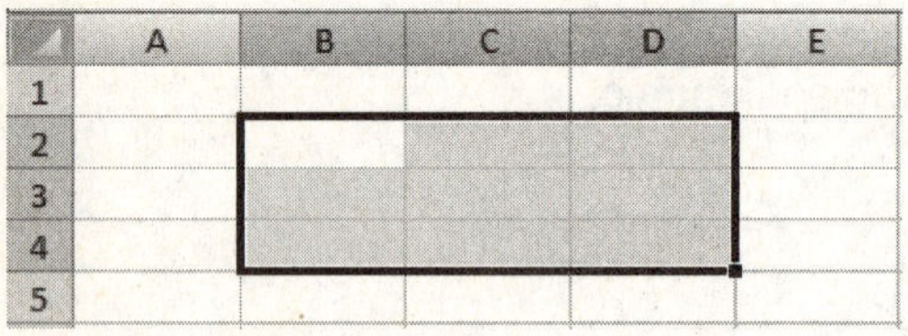

GS 3.2	Types of Information

Three types of information may be entered into an Excel worksheet.

1. **Text**. The term "text" refers to alphabetic characters or a combination of alphabetic characters and numbers, sometimes called "alphanumeric." The figure provides an example of an entry comprised solely of alphabetic characters (cell A1) and an entry comprised of a combination of alphabetic characters and numbers (cell B1).

	A	B
1	Sue Clark	25 years
2		

2. **Numeric**. Any cell entry comprised completely of numbers falls into the "numeric" category.

	A
1	2
2	9
3	8
4	5

3. **Formulas**. Formulas are a convenient way to perform mathematical operations on numbers already entered into the worksheet. Specific instructions are provided in this manual for problems that require the use of formulas.

GS 3.3	Entering Information

To enter information into a cell of the worksheet, first activate the cell. Then key in the desired information and press [**Enter**]. Pressing the [Enter] key moves you down to the next cell in that column. The information shown below was entered as follows:

1. Click in cell A1. Key in **1**. Press [**Enter**].

2. Key in **2**. Press [**Enter**].

3. Key in **3**. Press [**Enter**].

	A	B
1	1	
2	2	
3	3	
4		
5		

GS 3.4	Using Formulas

When you want to enter a formula, begin the cell entry with an equal sign (=). The arithmetic operators are displayed at the top of the next page.

Arithmetic operator	Meaning	Example
+	Addition	=3+2
-	Subtraction	=3-2
*	Multiplication	=3*2
/	Division	=3/2
^	Exponentiation	=3^2

Numbers, cell addresses, and functions can be used in formulas. For example, to sum the contents of cells A1 and B1, you can use the formula =A1+B1. To divide this sum by 2, you can use the formula =(A1+B1)/2. Note that Excel carries out expressions in parentheses first and then uses the results to complete the calculations of the formula. Formulas will sometimes not produce the desired results because parentheses were necessary but were not used.

Getting Started with Changing Information

GS 4.1	Editing Information in the Cells

There are several ways that you can edit information that has already been entered into a cell.

1. If you have not completed the entry, you can simply backspace and start over. Clicking on the red X to the left of the Formula Bar will also delete an incomplete cell entry.

2. If you have already completed the entry and another cell is activated, you can click in the cell you want to edit and then press either [**Delete**] or [**Backspace**] to clear the contents of the cell.

3. If you want to edit part of the information in a cell instead of deleting all of it, follow the instructions provided in the example.

- Let's say that you wanted to enter 1234 in cell A1 but instead entered 124. Return to cell **A1** to make it the active cell by either clicking on it with the mouse or by using the arrow keys.

- You will see A1 in the Name Box and 124 in the Formula Bar. Click between 2 and 4 in the Formula Bar so that the **I-beam** is positioned there. Enter the number 3 and press [**Enter**].

GS 4.2	Copying Information

To copy the information in one cell to another cell, follow these steps:

- First click on the source cell. Then, at the top of the screen, click the **Home** tab and select **Copy**.

- Click on the target cell where you want the information to be placed. Then, at the top of the screen, click **Home** and select **Paste**.

To copy a range of cells to another location in the worksheet, follow these steps:

- First click and drag over the range of cells that you want to copy so that they are highlighted. Then, at the top of the screen, click **Home** and select **Copy**.

- Click in the topmost cell of the target location. Then, at the top of the screen, click **Home** and select **Paste**.

To copy the contents of one cell to a range of cells follow these steps:

- Let's say that you have entered a formula in cell C1 that adds the contents of cells A1 and B1 and you would like to copy this formula to cells C2 and C3 so that C2 will contain the sum of A2 and B2 and cell C3 will contain the sum of A3 and B3.

- First click in cell C1 to make it the active cell. You will see =A1+B1 in the Formula Bar.

- At the top of the screen, click **Home** and select **Copy**.

- Highlight cells C2 and C3 by clicking and dragging over them.

- At the top of the screen, click **Home** and select **Paste**. The sums should now be displayed in cells C2 and C3.

| GS 4.3 | Moving Information |

If you would like to move the contents of one cell from one location to another in the worksheet, follow these steps:

- Click on the cell containing the information that you would like to move.

- At the top of the screen, click **Home** and select **Cut**.

- Click on the target cell where you want the information to be placed.

- At the top of the screen, click **Home** and select **Paste**.

If you would like to move the contents of a range of cells to a different location in the worksheet, follow these steps:

- Click and drag over the range of cells that you would like to move so that it is highlighted.

- At the top of the screen, click **Home** and select **Cut**.

- Click the topmost cell of the new location. (It is not necessary to click and drag over the entire range of the new location.)

- At the top of the screen, click **Home** and select **Paste**.

---

*If you make a mistake, just click the **Undo** arrow located in the upper-left corner of the screen. It looks like this:*

GS 4.4	Changing the Column Width

There are a couple of different ways that you can use to change the column width. Only one way will be described here. Output from the Descriptive Statistics data analysis tool will be used as an example. As you can see in the output displayed below, many of the labels in column A can only be partially viewed because the default column width is too narrow.

	A	B
1	*Test Score*	
2		
3	Mean	85.65
4	Standard I	2.274892
5	Median	85.5
6	Mode	80
7	Standard I	10.17362
8	Sample Vz	103.5026

Position the mouse pointer directly on the vertical line between A and B in the letter row at the top of the worksheet — | A | B | —so that it turns into a black plus sign.

Click and drag to the right until you can read all the output labels. (You can also click and drag to the left to make columns narrower.) After adjusting the column width, your output should appear similar to the output shown below.

	A	B
1	*Test Score*	
2		
3	Mean	85.65
4	Standard Error	2.274892
5	Median	85.5
6	Mode	80
7	Standard Deviation	10.17362
8	Sample Variance	103.5026

Getting Started with Sorting Information

| GS 5.1 | Sorting a Single Column of Information |

Let's say that you have entered "Score" in cell A1 and four numbers directly below it and that you would like to sort the numbers in ascending order.

- Click and drag from cell A1 to cell A5 so that the range of cells is highlighted.

You could also click directly on A *in the letter row at the top of the worksheet. This will result in all cells of column A being highlighted.*

	A
1	Score
2	15
3	79
4	18
5	2

- At the top of the screen, click the **Data** tab and select **Sort** in the Sort & Filter Group.

- In the Sort dialog box that appears, you are given the choice of sorting the information in column A in either Smallest to Largest or Largest to Smallest order. The **Smallest to Largest** order has already been selected. There is a checkmark in the box to the left of **My data has headers**. This means that the "Score" header will stay in cell A1 and will not be included in the sort. Click **OK** at the bottom of the dialog box.

Sort							? X
²₁ Add Level	✕ Delete Level	🖺 Copy Level	⬆	⬇	Options...		☑ My data has headers

Column	Sort On	Order
Sort by [Score ▼]	[Values ▼]	[Smallest to Largest ▼]

[OK] [Cancel]

The cells in column A should now be sorted in ascending order as shown below.

	A
1	Score
2	2
3	15
4	18
5	79

GS 5.2	Sorting Multiple Columns of Information

Your Excel data files will frequently contain multiple columns of information. When you sort multiple columns at the same time, Excel provides a number of options.

Let's say that you have a data file that contains the information shown below and that you would like to sort the file by GPA in descending order.

	A	B	C	D
1	Score	Age	Major	GPA
2	2	19	Music	3.1
3	15	19	History	2.4
4	18	22	English	2.7
5	79	20	English	3.7

- Click and drag from A1 down and across to D5 so that the entire range of cells is highlighted.

	A	B	C	D	E
1	Score	Age	Major	GPA	
2	2	19	Music	3.1	
3	15	19	History	2.4	
4	18	22	English	2.7	
5	79	20	English	3.7	
6					

- At the top of the screen, click the **Data** tab at the top of the screen and select **Sort** in the Sort & Filter group.

- In the Sort dialog box that appears, you are given the option of sorting the data by four different variables. You want to sort only by GPA in descending order. Click the down arrow to the right of the Sort by window until you see GPA and click on **GPA** to select it. Select the **Largest to Smallest** order. The checkmark in the box to the left of **My data has headers** means that the top row in the selected range will not be included in the sort, and the variable labels will stay in row 1. Click **OK**.

The sorted data file is shown below.

	A	B	C	D
1	Score	Age	Major	GPA
2	79	20	English	3.7
3	2	19	Music	3.1
4	18	22	English	2.7
5	15	19	History	2.4

Getting Started with Saving Information

GS 6.1	Saving Files

To save a newly created file for the first time, click the **Office icon** in the upper-left corner of the screen and select **Save**. A Save As dialog box will appear. You will need to select the location for saving the file by clicking on it.

The default file name, displayed in the File name window, is **Book1.** It is highly recommended you replace the default name with a name that is more descriptive.

Once you have saved a file, clicking the **Office icon** and **Save** will result in the file being saved in the same location under the same file name. No dialog box will appear. If you would like to save the file in a different location, you will need to select click the **Office icon** and select **Save As**.

GS 6.2	Naming Files

Windows and Mac versions of Excel will allow file names to have around 200 characters. You will find that long, descriptive names will be easier to work with than really short names. For example, if a file contains data that was collected in a survey of Milwaukee residents, you may want to name the file **Milwaukee resident survey** rather than **MRS.**

Several symbols cannot be used in file names. These include: forward slash (/), backslash (\), greater-than sign (>), less-than sign (<), asterisk (*), question mark (?), quotation mark ("), pipe symbol (|), color (:), and semicolon (;).

Getting Started with Printing Information

GS 7.1	Printing Worksheets

To print a worksheet, click the **Office icon** in the upper-left corner of the screen and select **Print**. The Print dialog box will appear.

Under Print range, you will usually select **All**, and under Print what, you will usually select **Active sheet(s)**. The default number of copies is 1, but you can increase this if you need more copies. When the Print dialog box has been completed as you would like, click **OK**.

GS 7.2	Page Setup

Excel provides a number of page setup options for printing worksheets. To access these options, click the **Page Layout** tab at the top of the screen and select from the command groups. In **Page Setup**, you may want to select **Orientation** to change from portrait to landscape orientation for worksheets that have several columns of data. Under the **Gridlines** command in the **Sheet Options** groups, you may want to select **Print** so that gridlines appear in your hard copies.

Getting Started with Add-ins

GS 8.1	Loading Excel's Analysis Toolpak

The Analysis Toolpak is an Excel Add-In that may not necessarily be loaded. If it does not appear in the Analysis group of the Data ribbon as shown below, then you will need to load it.

First click the **Office icon** and, at the bottom of the dialog box, select **Excel Options**. Select **Add-Ins** from the list on the left. Analysis ToolPak and Analysis ToolPak – VBA should both be in the list of Active Application Add-Ins. If you need to add one or both of them, first click on the name to select it. Then click **Go** at the bottom of the screen. In the Add-Ins dialog box, place a checkmark in the box next to the add-in to make it active. Click **OK**.

GS 8.2	Loading the DDXL Add-In

Data Desk/XL (DDXL) is a statistical add-in that is included on the CD-ROM that accompanies this manual. The instructions that are given here also appear in the DDXL Read Me.txt file.

DDXL Student version 2.2.1 works on Windows 2000/XP/Vista in Excel 2000 or better and in Excel 2007.

To use the DDXL Microsoft Excel add-in, you first need to install DDXL on your computer. Find Install_DDXL_Student.exe in the DDXL folder on the CD-ROM. Double-click to launch. Click Finish when prompted.

After you have installed DDXL successfully, you must setup the add-in within Excel itself. To do that, use the following instructions:

To Install the DDXL Add-in in Excel 2000 or Later:

1) Launch Microsoft Excel 2000 or later. Make certain a worksheet is open.
2) From the Tools menu, select Add-Ins...
3) Choose Browse...
4) Navigate to the file "DDXL Add-In.xll" and click OK.
      (The default location is C:\Program Files\DDXL\DDXL Add-In.xll)
5) Verify that there is a checkmark in front of the DDXL Add-In, click OK.

NOTE: This version of DDXL does not require you to enter an Activation Key.

To Install the DDXL Add-in in Excel 2007:

1) Launch Microsoft Excel 2007. Make certain a worksheet is open.
2) Click the Microsoft Office Button.
3) Click the Excel Options button.
4) Click the Add-Ins listing to select, make certain Excel Add-ins is listed in the Manage area. If not, select from the drop-down menu.
5) Click Go.
6) In the Add-Ins box click Browse...
7) Navigate to the file "DDXL Add-In.xll" and click OK.
      (The default location is C:\Program Files\DDXL\DDXL Add-In.xll)
8) Verify that there is a checkmark in front of the DDXL Add-In, click OK.

NOTE: This version of DDXL does not require you to enter an Activation Key.

When you click the Add-Ins tab at the top of your screen, you will see DDXL in the Menu Commands group.

When you click **DDXL**, you will see the DDXL menu choices.

Introduction to Statistics

Technology

▶ Example (pg. 34)	Generating a List of Random Numbers

You will be generating a list of 15 random numbers between 1 and 167 to use in selecting a random sample of 15 cars assembled at an auto plant.

1. Open a new Excel worksheet.

2. At the top of the screen, click **Formulas** and select **Insert Function**.

3. Select the **All** category and the **RANDBETWEEN** function. Click **OK**.

21

4. Complete the RANDBETWEEN dialog box as shown below. The function will return one random integer between 1 and 167. Click **OK**.

Function Arguments		?	X

RANDBETWEEN

Bottom	1	[icon]	= 1
Top	167	[icon]	= 167

= Volatile

Returns a random number between the numbers you specify.

  Top is the largest integer RANDBETWEEN will return.

Formula result = Volatile

Help on this function   OK   Cancel

5. The output is displayed below. Because the number was generated randomly, it is not likely that your output will be exactly the same.

	A
1	71

6. To obtain 15 randomly selected numbers, copy the contents of cell A1 to cells A2:A15. You might want to generate more than 15 random numbers because there is a chance that your output will contain repetitions. In addition, you will notice that the number in cell A1 changes when you copy the RANDBETWEEN function.

	A
1	75
2	4
3	146
4	7
5	94
6	8
7	85
8	161
9	109
10	88
11	84
12	125
13	76
14	8
15	94

◀

▶ Exercise 1 (pg. 35)	Generating a List of Random Numbers for the SEC

You will be generating a list of 10 random numbers between 1 and 86 to use in selecting a random sample of brokers. You will also be ordering the generated list from lowest to highest.

If Data Analysis does not appear as a choice in the Data ribbon, you will need to load the Microsoft Excel Analysis ToolPak add-in. Follow the procedure in Section GS 8.1 before continuing.

1. Open a new Excel worksheet.

2. Use the Fill command to enter the numbers 1 to 86 in column A. To do this, begin by entering **1** in cell **A1**. Then click in cell **A1**.

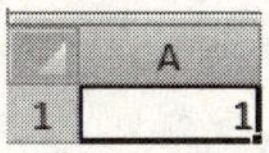

	A
1	1

3. At the top of the screen, click **Home** and select **Fill** in the Editing group.

4. Select **Series**. Complete the Series dialog box as shown below. Click **OK**.

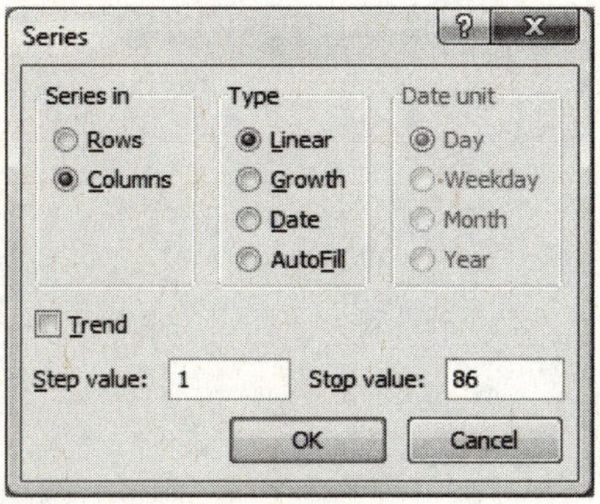

Series

Series in	Type	Date unit
○ Rows	● Linear	● Day
● Columns	○ Growth	○ Weekday
	○ Date	○ Month
	○ AutoFill	○ Year

☐ Trend

Step value: 1   Stop value: 86

OK   Cancel

5. Column A of your worksheet should now display the numbers 1 to 86. At the top of the screen, click **Data** and select **Data Analysis** in the Analysis group. Select **Sampling** and click **OK**.

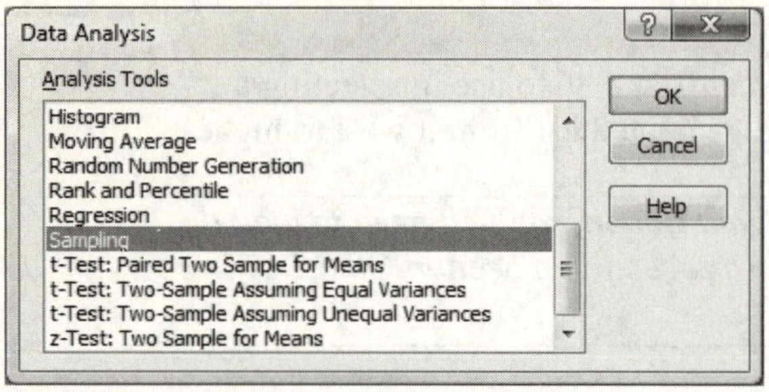

6. Complete the Sampling dialog box as shown below. The output will be displayed in the current worksheet starting in cell C1. Click **OK**.

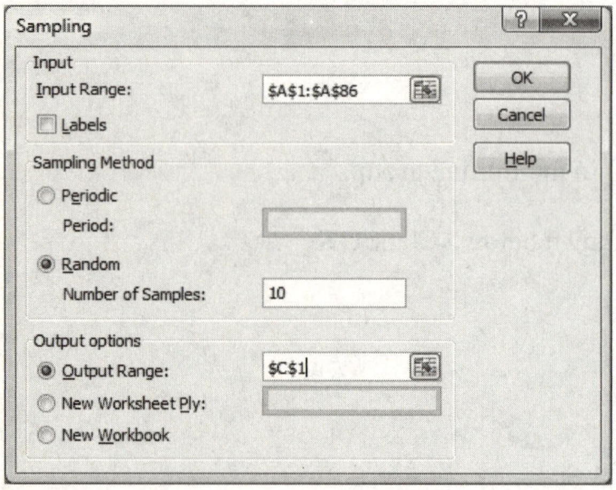

7. The output is displayed below. Because the numbers were generated randomly, it is not
    likely that your output will be exactly the same.

*If your output contains repetitions, you may want to generate a new sample or generate more
than 10 numbers.*

	A	B	C
1	1		48
2	2		29
3	3		38
4	4		46
5	5		81
6	6		84
7	7		13
8	8		5
9	9		82
10	10		66

8. Sort the numbers in ascending order.

For instructions on how to sort, refer to Sections GS 5.1 and GS 5.2.

	A	B	C
1	1		5
2	2		13
3	3		29
4	4		38
5	5		46
6	6		48
7	7		66
8	8		81
9	9		82
10	10		84

◀

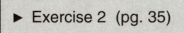 ► Exercise 2 (pg. 35)   Generating a List of Random Numbers for a Quality
Control Department

You will be generating a list of 25 random numbers between 1 and 300 and then ordering the
generated list from lowest to highest.

If Data Analysis does not appear as a choice in the Data ribbon, you will need to load the Microsoft Excel Analysis ToolPak add-in. Follow the procedure in Section GS 8.1 before continuing.

---

1.   Open a new Excel worksheet.

2.   Use the Fill command to enter the numbers 1 to 300 in column A. To do this, begin by entering **1** in cell **A1**. Then click in cell **A1**.

A
1

3.   At the top of the screen, click **Home** and select **Fill** in the Editing group.

4.   Select **Series**. Complete the Series dialog box as shown below. Click **OK**.

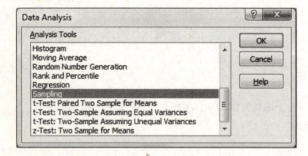

5.   Column A of your worksheet should now display the numbers 1 to 300. At the top of the screen, click **Data** and select **Data Analysis** in the Analysis group. Select **Sampling** and click **OK**.

6. Complete the Sampling dialog box as shown below. The output will be displayed in the current worksheet starting in cell C1. Click **OK**.

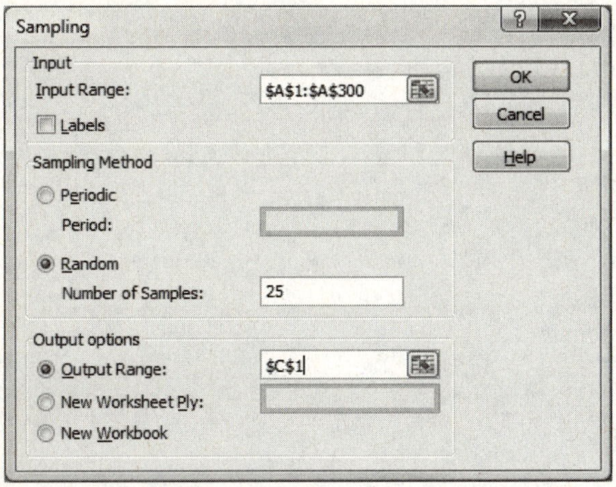

If your output contains repetitions, you may want to generate a new sample or generate more than 25 numbers.

4. Sort the numbers in ascending order.

For instructions on how to sort, see Sections GS 5.1 and GS 5.2.

The sorted set of 25 random numbers is displayed below. Because the numbers were generated randomly, it is unlikely that your output will be exactly the same.

	A	B	C
1	1		21
2	2		28
3	3		34
4	4		42
5	5		65
6	6		70
7	7		78
8	8		93
9	9		94
10	10		99
11	11		103
12	12		107
13	13		114
14	14		143
15	15		158
16	16		187
17	17		197
18	18		206
19	19		214
20	20		224
21	21		225
22	22		277
23	23		287
24	24		288
25	25		300

◄

| ► Exercise 3 (pg. 35) | Generating Three Random Samples from a Population of Ten Digits |

You will be generating three random samples of five digits each from the population: 0, 1, 2, 3, 4, 5, 6, 7, 8, 9. You will also compute the average of each sample.

If Data Analysis does not appear as a choice in the Data ribbon, you will need to load the Microsoft Excel Analysis ToolPak add-in. Follow the procedure in Section GS 8.1 before continuing.

1. Open a new Excel worksheet. Enter the numbers 0 through 9 in column A.

	A
1	0
2	1
3	2
4	3
5	4
6	5
7	6
8	7
9	8
10	9

2. Compute the average using the AVERAGE function. To do this, first click in the cell immediately below the last number in the population, cell **A11**, where you will place the average. Then, at the top of the screen click **Formulas** and select **Insert Function**.

3. Select the **Statistical** category. Select the **AVERAGE** function. Click **OK**.

4. Complete the AVERAGE dialog box as shown at the top of the next page. A1:A10 is the worksheet location of the population of ten digits. Click **OK**.

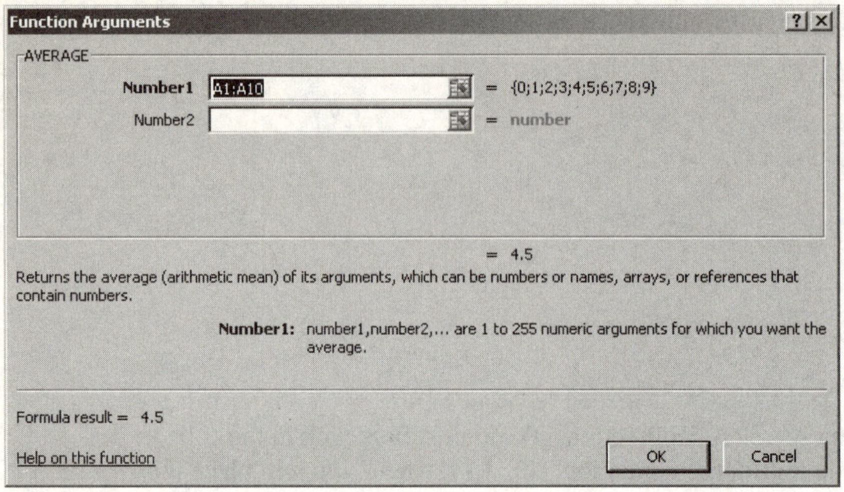

The average, 4.5, is displayed in cell A11 of the worksheet.

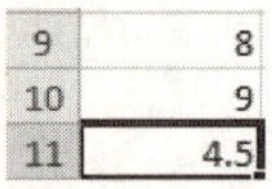

9	8
10	9
11	4.5

7. You will now select the first random sample of five digits. At the top of the screen, click **Data** and select **Data Analysis** in the Analysis group. Select **Sampling** and click **OK**.

8. Complete the Sampling dialog box as shown below. The output will be displayed in the current worksheet starting in cell C1. Click **OK**

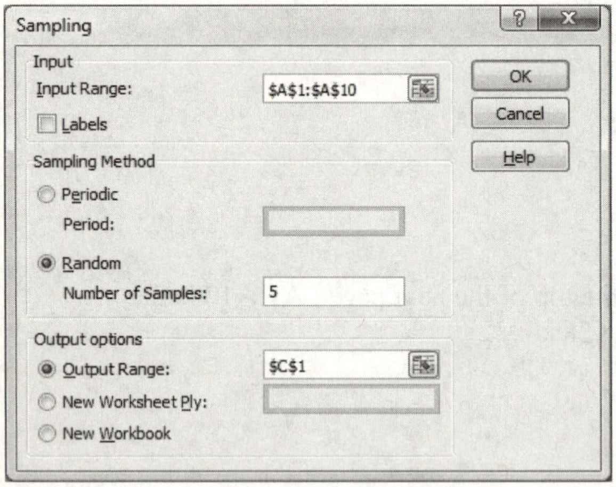

9. Use the AVERAGE function to find the average. To do this, first click in the cell immediately below the last number in the sample—cell **A6**. Then, follow steps 3 and 4 above using the sample data range C1:C5.

10. The average of the sample, 5.2, is now displayed in cell A6 of the worksheet. Because the numbers were generated randomly, it is not likely that your result will be exactly the same. Repeat steps 7 to 9 to generate two more samples and compute the averages.

	A	B	C
1	0		7
2	1		5
3	2		6
4	3		0
5	4		8
6	5		5.2
7	6		
8	7		
9	8		
10	9		
11	4.5		

◄

► Exercise 5 (pg. 35)   Simulating Rolling a Six-Sided Die 60 Times

You will be simulating rolling a six-sided die 60 times and making a tally of the results.

If Data Analysis does not appear as a choice in the Data ribbon, you will need to load the Microsoft Excel Analysis ToolPak add-in. Follow the procedure in Section GS 8.1 before continuing.

If the DDXL add-in has not been loaded, you will need to load it before continuing. Follow the instructions in Section GS 8.2.

1. Open a new Excel worksheet and enter the digits 1 through 6 in column A. These digits represent the possible outcomes of rolling a die. Type **Outcome** in C1. "Outcome" will be the variable name you will use in the DDXL's Bar Chart procedure.

	A	B	C
1	1		Outcome
2	2		
3	3		
4	4		
5	5		
6	6		

2. At the top of the screen, click **Data** and select **Data Analysis**. Select **Sampling** and click **OK**.

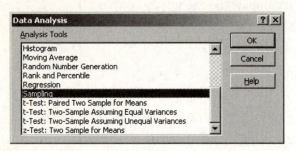

3. Complete the Sampling dialog box as shown below. Click **OK**.

The output will be displayed in the current worksheet starting in cell C2. The first 7 entries are shown at the top of the next page. Because the numbers were generated randomly, it is not likely that your output will be exactly the same.

	A	B	C
1	1		Outcome
2	2		3
3	3		6
4	4		5
5	5		1
6	6		6
7			5

4. To make a tally of the results, first highlight the range of the outcomes, **C1:C61**. Click **Add-Ins** and select **DDXL**. Select **Charts and Plots**.

5. Click the down arrow to the right of **CLICK ME** and select **Bar Chart**.

6. Click on **Outcome** in the window under Names and Columns and then click the arrow to the right of the window under Categorical Variable so that Outcome now appears in the Categorical Variable window. Click **OK**.

The output is shown at the top of the next page. The bar chart output includes a chart and a frequency table. Because the numbers were generated randomly, it is not likely that your output will be exactly the same.

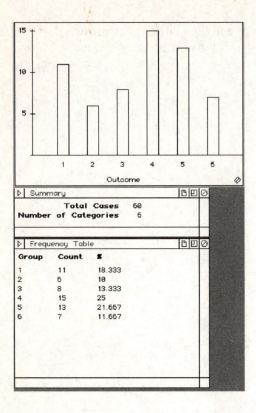

Descriptive Statistics

Section 2.1 Frequency Distributions and Their Graphs

| ▶ Example 7 (pg. 46) | Using Technology to Construct Histograms |

You will be constructing a histogram for the frequency distribution of the GPS navigator data in Example 1 on page 39. You will be using one of Excel's data analysis tools.

If Data Analysis does not appear as a choice in the Data ribbon, you will need to load the Microsoft Excel Analysis ToolPak add-in. Follow the procedure in Section GS 8.1 before continuing.

1. Open worksheet "GPS" in the Chapter 2 folder. These data represent the prices (in dollars) of 30 GPS navigators.

2. Excel's histogram tool uses grouped data to generate a frequency distribution and a frequency histogram. The procedure requires that you indicate a "bin" for each class. The number that you specify for each bin is actually the upper limit of the class. The upper limits for the GPS data are given on page 39 of your text and are based on a class width of 56. You see that 114 is the upper limit for the first class, 170 is the upper limit for the second class, and so on. The bin for the first class will contain a count of all observations less than or equal to 114, the bin for the second class will contain a count of all observations between 115 and 170, and so on. Enter **Bin** in cell B1 and key in the upper limits in column B as shown below.

	A	B
1	GPS Prices	Bin
2	90	114
3	130	170
4	400	226
5	200	282
6	350	338
7	70	394
8	325	450

35

3.   Click **Data** and select **Data Analysis**. Select **Histogram** and click **OK**.

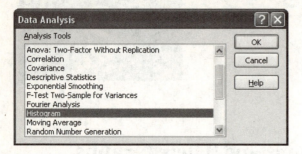

4.   Complete the Histogram dialog box as shown below. Be sure to click **Labels** and **Chart Output**. Click **OK**.

To enter the input and bin ranges quickly, follow these steps. First click in the Input Range field of the dialog box. Then, in the worksheet, click and drag from cell A1 to cell A31. You will then see $A$1:$A$31 displayed in the Input Range field. Next click in the Bin Range field. Then click and drag from cell B1 through cell B8 in the worksheet. You will see $B$1:$B$8 displayed in the Bin Range field.

It is necessary to put a checkmark in the box to the left of Labels. In your worksheet, "GPS Prices" appears in cell A1, and "Bin" appears in cell B1. Because you included these cells in the Input Range and Bin Range, respectively, you need to let Excel know that these cells contain labels rather than data. Otherwise Excel will attempt to use the information in these cells when constructing the frequency distribution and histogram.

You should see output similar to the output displayed below.

	A	B	C	D	E	F	G	H	I
1	Bin	Frequency							
2	114	5							
3	170	8							
4	226	6							
5	282	5							
6	338	2							
7	394	1							
8	450	3							
9	More	0							
10									

If you want to change the height and width of the chart, begin by clicking anywhere within the figure. Handles appear along the perimeter. You can change the shape of the figure by clicking on a handle and dragging it. You can make the figure small and short or you can make it tall and wide. You can also click within the figure and drag it to move it to a different location on the worksheet.

5. You will now follow steps to modify the histogram so that it is displayed in a more informative and attractive manner. First, make the chart taller and wider so that it is easier to read. To do this, first click within the figure near a border. Dotted handles appear. Click on the center handle on the bottom border of the figure and drag it down a few rows. Then click the handle on the right side and drag it over about an inch.

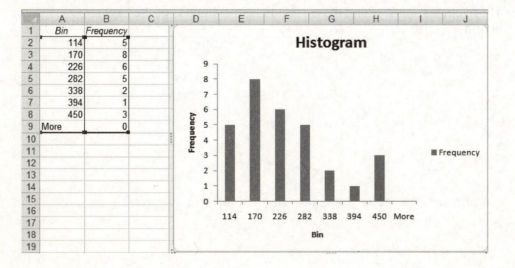

6. Next remove the space between the vertical bars. **Right-click** on one of the vertical bars. Select **Format Data Series** from the shortcut menu that appears.

7. Move the Gap Width arrow all the way to the right for a gap width of 0%. Click **Close**.

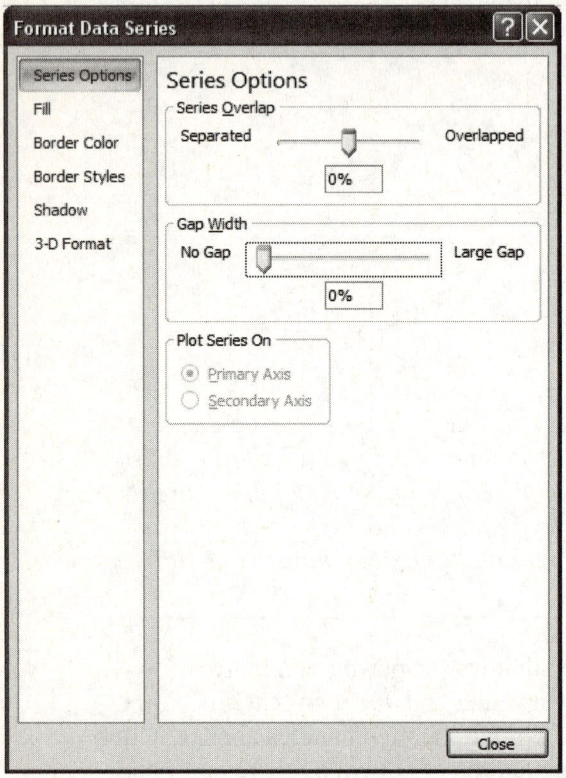

```
Format Data Series                            [?][X]

  Series Options      Series Options
  Fill                  ┌ Series Overlap ──────────────────────┐
  Border Color          │  Separated        ───┬───  Overlapped │
  Border Styles         │                   [0%]                 │
  Shadow                └──────────────────────────────────────┘
  3-D Format            ┌ Gap Width ───────────────────────────┐
                        │  No Gap  ─┬─────────────  Large Gap    │
                        │          [0%]                          │
                        └──────────────────────────────────────┘
                        ┌ Plot Series On ──────────────────────┐
                        │  ⦿ Primary Axis                        │
                        │  ○ Secondary Axis                      │
                        └──────────────────────────────────────┘

                                              [ Close ]
```

8. Next, change the X-axis values from upper limits to midpoints. The midpoints are displayed in a table on page 41 of your textbook. Enter these midpoints in column C of the Excel worksheet as shown below.

	A	B	C
1	*Bin*	*Frequency*	Midpoint
2	114	5	86.5
3	170	8	142.5
4	226	6	198.5
5	282	5	254.5
6	338	2	310.5
7	394	1	366.5
8	450	3	422.5
9	More	0	

9. **Right-click** on a vertical bar. Click **Select Data** in the shortcut menu that appears.

10. Click **Edit** under Horizontal (Category) Axis Labels. In the Excel worksheet, click and drag over the range **C2:C8** to enter the location of the midpoints into the Axis Labels dialog box. Click **OK**.

Axis Labels	? X
Axis label range:	
=Sheet4!$C$2:$C$8 ▥	= 86.5, 142.5, 1...
OK Cancel	

11. Click **OK** at the bottom of the Select Data Source dialog box.

Select Data Source   ? X

Chart data range: ▥

The data range is too complex to be displayed. If a new range is selected, it will replace all of the series in the Series panel.

↓   ▣ Switch Row/Column   ↓

Legend Entries (Series)   Horizontal (Category) Axis Labels

⊞ Add   ✎ Edit   ✗ Remove   ⬆ ⬇   ✎ Edit

Frequency

86.5
142.5
198.5
254.5
310.5

Hidden and Empty Cells   OK   Cancel

12. Next, you will change the chart title and the x-axis title. Click on the word "Histogram" so that it appears in a box. Change the title to **Prices of GPS Navigators**. Click on "Bin" and change the x-axis title to **Price (in dollars)**.

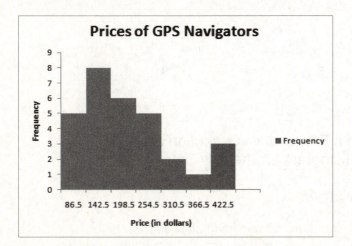

13. Now you will add gridlines. **Right-click** on any number in the y-axis scale. I clicked on 8. Select **Add Major Gridlines** from the menu that appears.

14. To remove the legend, first click directly on **Frequency** at the right side of the chart so that it is displayed within a border. Then press [**Delete**].

The completed chart is shown below.

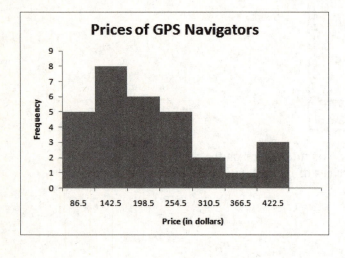

▶ Exercise 31 (pg. 50)   Constructing a Frequency Distribution and Frequency Histogram for the July Sales Data

1. Open worksheet "Ex2_1-31" in the Chapter 2 folder.

2. Sort the data so that it is easy to identify the minimum and maximum data entries. In the sorted data set, you see that the minimum sales value is 1000 and that the maximum is 7119.

For instructions on how to sort, see Sections GS 5.1 and GS 5.2.

---

3. Calculate the class width using the formula given in your textbook:

$$\text{Class width} = \frac{\text{Maximum data entry} - \text{Minimum data entry}}{\text{Number of classes}}$$

The exercise instructs you to use 6 for the number of classes.

$$\text{Class width} = \frac{7119 - 1000}{6} = 1019.83$$

Round up to 1020.

4. The textbook instructs you to use the minimum data entry as the lower limit of the first class. To find the remaining lower limits, add the class width of 1020 to the lower limit of each previous class. To do this, begin by entering **Lower Limit** in cell B1 of the worksheet and **1000** in cell B2. You will now use a formula to compute the remaining lower limits, and you will have Excel do these calculations for you. In cell B3, enter the formula **=B2+1020** as shown below. Press **[Enter]**.

	A	B	C
1	July sales	Lower Limit	
2	1000	1000	
3	1030	=B2+1020	

5. Click in cell **B3** where 2020 now appears and copy the formula in cell B3 to cells B4 through B8. Because 7119 is the maximum data entry, you don't need 7120 for the histogram. However, you will be using 7120 when calculating the upper limit for the last class interval.

	A	B	C
1	July sales	Lower Limit	
2	1000	1000	
3	1030	2020	
4	1077	3040	
5	1355	4060	
6	1500	5080	
7	1512	6100	
8	1643	7120	

6. Enter **Upper Limit** in cell C1 of the worksheet. The textbook says that the upper limit is equal to one less than the lower limit of the next higher class. You will use a formula to compute the upper limits, and you will have Excel do the computations for you. Click in cell **C2** and enter the formula **=B3-1** as shown below. Press **[Enter]**.

	A	B	C	D
1	July sales	Lower Limit	Upper Limit	
2	1000	1000	=B3-1	
3	1030	2020		

7. Copy the formula in cell C2 (where 2019 now appears) to cells C3 through C7. These are the upper limits you will use as bins for constructing the histogram chart.

	A	B	C	D
1	July sales	Lower Limit	Upper Limit	
2	1000	1000	2019	
3	1030	2020	3039	
4	1077	3040	4059	
5	1355	4060	5079	
6	1500	5080	6099	
7	1512	6100	7119	

8. Click **Data** and select **Data Analysis**. Select **Histogram** and click **OK**.

If Data Analysis does not appear as a choice in the Data ribbon, you will need to load the Microsoft Excel Analysis ToolPak add-in. Follow the procedure in Section GS 8.1 before continuing.

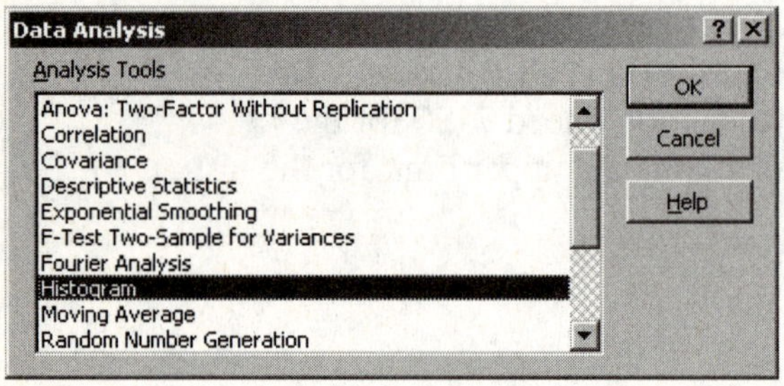

Data Analysis [? X]

Analysis Tools

Anova: Two-Factor Without Replication
Correlation
Covariance
Descriptive Statistics
Exponential Smoothing
F-Test Two-Sample for Variances
Fourier Analysis
Histogram
Moving Average
Random Number Generation

OK
Cancel
Help

9. Fill in the fields in the Histogram dialog box as shown below. The July sales data are located in cells A1 through A23 of the worksheet. The bins (upper limits) are located in cells C1 through C7. The top cells in these ranges are labels ("July sales" and "Upper Limit"), so click in the Labels box to place a checkmark there. The output will be placed in a new Excel workbook. The output will include a chart. Click **OK**.

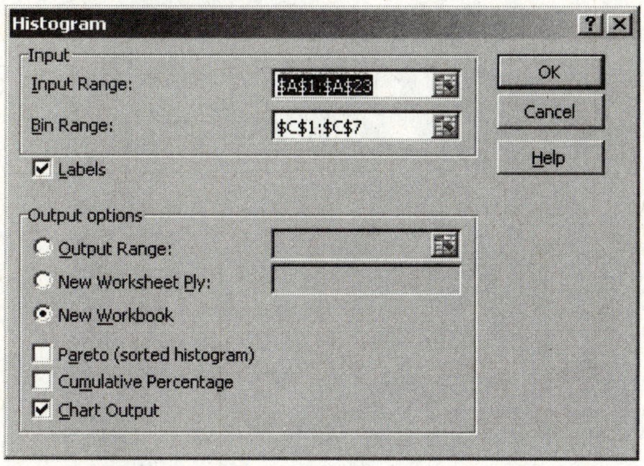

Histogram	? X
Input	
Input Range: $A$1:$A$23	OK
Bin Range: $C$1:$C$7	Cancel
☑ Labels	Help
Output options	
○ Output Range:	
○ New Worksheet Ply:	
● New Workbook	
☐ Pareto (sorted histogram)	
☐ Cumulative Percentage	
☑ Chart Output	

You should see output similar to the output displayed below.

	A	B	C	D	E	F	G	H	I	J
1	Upper Limit	Frequency								
2	2019	12					Histogram			
3	3039	3								
4	4059	2								
5	5079	3								
6	6099	1							■ Frequency	
7	7119	1								
8	More	0				2019 3039 4059 5079 6099 7119 More				
9										
10					Upper Limit					
11										

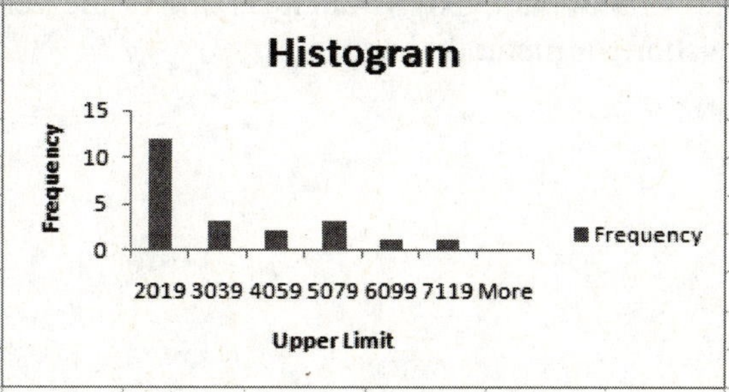

10. You will now follow steps to modify the histogram so that it is displayed in a more accurate and informative manner. First, make the chart taller so that it is easier to read. To do this, first click within the figure near a border. Dotted handles appear. Click on the center handle on the bottom border of the figure and drag it down a few rows. Next, click on the handle on the right side and drag it a couple of columns to the right.

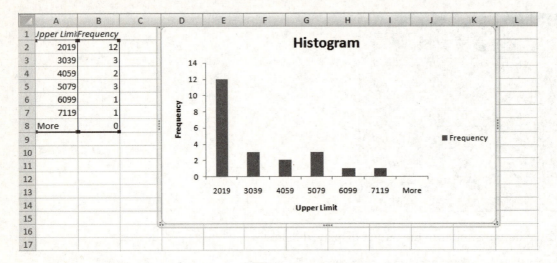

11. Remove the space between the vertical bars. **Right-click** on one of the vertical bars. Select **Format Data Series** from the shortcut menu that appears.

12. Move the Gap Width arrow all the way to **No Gap**. Click **Close**.

13. Delete the word "More" from the X axis. To do this, **right-click** on a vertical bar and select **Select Data** from the shortcut menu that appears.

14. Click **Edit** under Legend Entries (Series).

Select Data Source ? ✕

Chart data range: `=Sheet1!$A$2:$B$8`

⟶ 🔲 Switch Row/Column ⟶

Legend Entries (Series)
📋 Add   📝 Edit   ✕ Remove   ⬆ ⬇

Frequency

Horizontal (Category) Axis Labels
📝 Edit
2019
3039
4059
5079
6099

Hidden and Empty Cells   OK   Cancel

15. You do not want to include row 8 of the frequency distribution because that is the row containing information related to the "More" category. Change the 8 to 7 in the **Series values** field so that it reads **=Sheet1!$B$2:$B$7**. Click **OK**. Click **OK** in the Select Data Source dialog box.

Edit Series ? ✕

Series name:
[   ] 🔲 Select Range

Series values:
[=Sheet1!$B$2:$B$7] 🔲 = 12, 3, 2, 3, 1...

          OK   Cancel

16. You will now revise the chart and axis titles. Click on the "Histogram" chart title and change it to **July Sales**. Click on "Upper Limit" and change it to **Dollars**.

17. To remove the legend, click on the **Frequency** legend at the right and press [**Delete**].

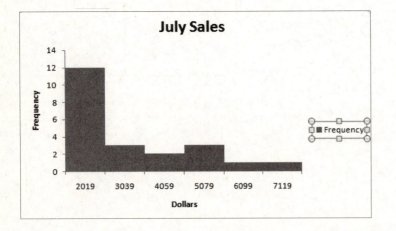

18.   To add gridlines, **right-click** on a number on the Y-axis scale. I clicked on 8.   Select **Add Major Gridlines**.

The completed histogram is shown below.

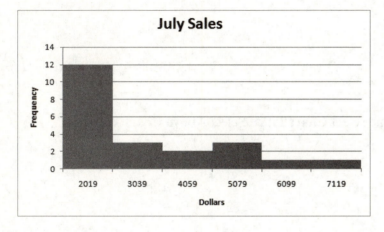

Section 2.2 More Graphs and Displays

► Example 4 (pg. 56)   Constructing a Pie Chart
_____

1.   Open a new Excel worksheet and enter the earned degrees frequency table as shown at the top of the next page.

	A	B
1	Type of degree	Number (in thousands)
2	Associate's	455
3	Bachelor's	1052
4	Master's	325
5	First professional	71
6	Doctoral	38

2. Highlight cells **A2:B6**. Then click **Insert** at the top of the screen and select **Pie** in the Charts group. Select the leftmost diagram in the top row.

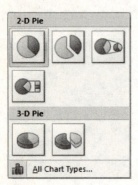

3. Click **Layout** at the top of the screen and select **Chart Title**. Select **Above Chart**.

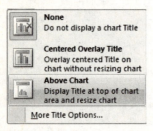

4. Replace "Chart Title" with **Earned Degrees Conferred in 1990**.

5. **Right-click** directly on the pie chart and select **Add Data Labels**.

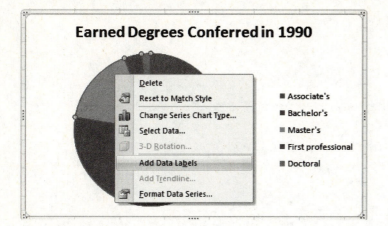

6. **Right-click** on a data label. I clicked on 604. Select **Format Data Labels**.

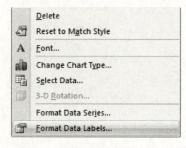

7.   Select **Percentage** and **Show Leader Lines**.   Click **Close**.

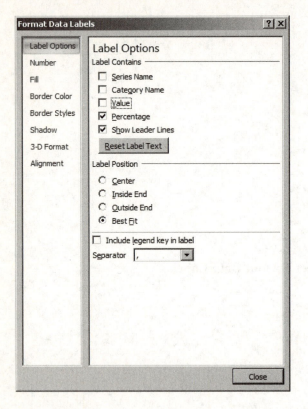

8.   You will now move the pie chart to a new worksheet. **Right-click** in the white space around the chart.   Select **Move Chart**.

9. Select **New sheet**. Click **OK**.

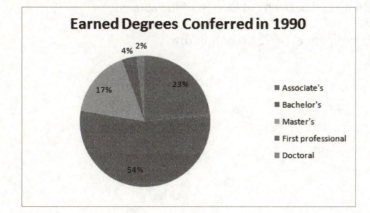

The completed pie chart is shown below.

Earned Degrees Conferred in 1990

- ■ Associate's
- ■ Bachelor's
- ■ Master's
- ■ First professional
- ■ Doctoral

2%
4%
17%
23%
54%

◄

▶ Example 5 (pg. 57)   Constructing a Pareto Chart

1. Open a new Excel worksheet, and enter the customer complaint information as shown below.

	A	B
1	Source	Frequency
2	Home furnishing	7792
3	Computer sales & service	5733
4	Auto dealers	14668
5	Auto repair	9728
6	Dry cleaning	4649

2. In a Pareto chart, the vertical bars are placed in order of decreasing height. Excel's Chart function will display the information in the order in which it appears in the worksheet. So, you first need to sort the information in descending order by Frequency.

For instructions on how to sort, see Sections GS 5.1 and GS 5.2.

	A	B
1	Source	Frequency
2	Auto dealers	14668
3	Auto repair	9728
4	Home furnishing	7792
5	Computer sales & service	5733
6	Dry cleaning	4649

3. After sorting the data, highlight cells **A1:B6** and click **Insert** at the top of the screen.

4. Select a **Column** chart and select the leftmost figure in the 2-D column row.

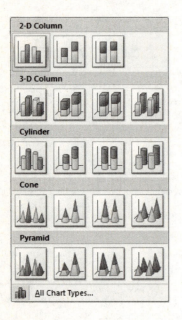

5. Next you will add an X-axis title. Click **Layout** at the top of the screen and select **Axis Titles→Primary Horizontal Axis→Title Below Axis**. Replace "Axis Title" with **Type of Business**.

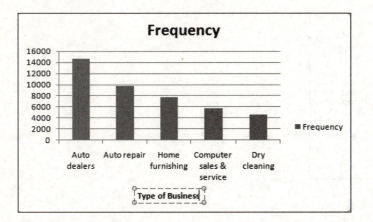

6. Now you will add the Y-axis title. Click **Layout** at the top of the screen and select **Axis Titles→Primary Vertical Axis Title →Vertical Title**. Replace "Axis Title" with **Frequency**.

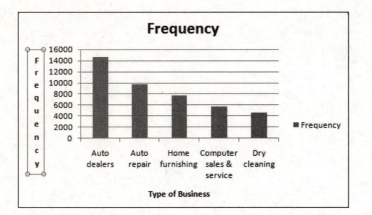

7. To remove the legend, click on the **Frequency** legend at the right and press [**Delete**].

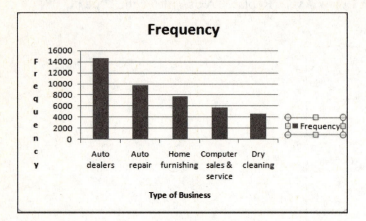

Your chart should look similar to the one shown below.

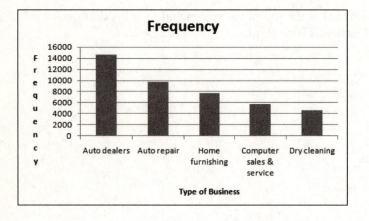

◀

▶ Example 7 (pg. 59)   Constructing a Time Series Chart

1. Open worksheet "Cellular Phones" in the Chapter 2 folder.

2. Highlight cells **A2:B12**. At the top of the screen, click **Insert** and select **Scatter** in the Charts group.

	A	B	C
1	Year	Subscribers	Avg. Bill $
2	1998	69.2	39.43
3	1999	86.0	41.24
4	2000	109.5	45.27
5	2001	128.4	47.37
6	2002	140.8	48.40
7	2003	158.7	49.91
8	2004	182.1	50.64
9	2005	207.9	49.98
10	2006	233.0	50.56
11	2007	255.4	49.79
12	2008	270.3	50.07

3. Select the leftmost scatter diagram in the top row.

4. At the top of the screen, click **Design** and select **Layout 1** in the Charts Layout group so that you can add a chart title and axis titles.

5. Replace "Chart Title" with **Cellular Telephone Subscribers**. Replace the Y-Axis Title with **Subscribers (in millions)**. Replace the X-Axis Title with **Year**.

6. To remove the legend, click on the **Series1** legend to the right of the chart. Press [**Delete**].

Your scatter plot should now look similar to the one shown at the top of the next page.

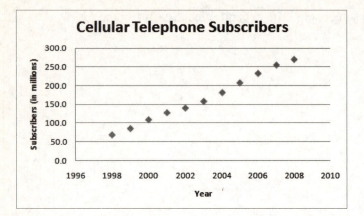

Section 2.3 Measures of Central Tendency

| ▶ Example 6 (pg. 68) | Comparing the Mean, Median, and Mode |

If Data Analysis does not appear as a choice in the Data ribbon, you will need to load the Microsoft Excel Analysis ToolPak add-in. Follow the procedure in Section GS 8.1 before continuing.

1. Open worksheet "Ages" in the Chapter 2 folder. The instructions for Try It Yourself 6 at the bottom of page 68 tell you to remove the data entry of 65 years. Because 65 is the bottom entry in the data set, you do not need to delete it. Instead, you will not include it in the input range.

2. At the top of the screen, click **Data** and select **Data Analysis**. Select **Descriptive Statistics** and click **OK**.

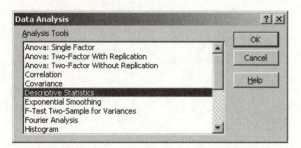

3. Complete the Descriptive Statistics dialog box as shown below. The entry in the **Input Range** field is the worksheet location of the age data. The checkmark to the left of **Labels in First Row** lets Excel know that the entry in cell A1 is a label and is not to be included in the computations. The output will be placed in a **New Worksheet**. The checkmark to the left of **Summary statistics** requests that the output include summary statistics for the specified set of data. Click **OK**.

4. You will want to increase the width of column A of the output so that you can read the labels. Your output should be similar to the output shown at the top of the next page. The mean of the sample is 21.58, the median is 21, and the mode is 20.

Be careful when using the value of the mode reported in the Descriptive Statistics output. If there is a tie for the mode, Excel reports only the first modal value that occurs in the data set. Therefore, it is always a good idea to construct a frequency distribution table to go along with the Descriptive Statistics output.

	A	B
1	*Ages of students in class*	
2		
3	Mean	21.57895
4	Standard Error	0.327276
5	Median	21
6	Mode	20
7	Standard Deviation	1.426565
8	Sample Variance	2.035088
9	Kurtosis	-1.27106
10	Skewness	0.335563
11	Range	4
12	Minimum	20
13	Maximum	24
14	Sum	410
15	Count	19

5. Construct a frequency distribution table to see if the data set is unimodal or multimodal. Go back to the worksheet containing the age data. To do this, click on the **Sheet1** tab near the bottom of the screen.

If the DDXL add-in has not been loaded, you will need to load it before continuing. Follow the instructions in Section GS 8.2.

6. Click and drag over the age data range, A1:A20.

7. Click **Add-Ins** at the top of the screen and select **DDXL**. Select **Charts and Plots**.

8. Click the down arrow under Function type and select **Histogram**.

Charts and Plots Dialog

Function type:

** CLICK ME **

** CLICK ME **
3D Rotating Plot
Bar Chart
Boxplot
Boxplot by Groups
Color Scatterplot
Dotplot
Dotplot by Groups
Histogram
Normal Probability Plot
Pie Chart
Scatterplot
StackedDotplot

9. Select **Ages of students in class** in the Names and Columns window. Then click the arrow under Quantitative Variable on the left to select Ages of students in class for the histogram. Note that a checkmark has already been placed in the box next to **First row is variable names**. Click **OK**.

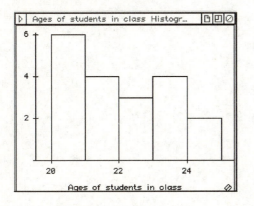

10. The chart in the output shows you that the age distribution is unimodal. The modal value of 20 has a frequency of six.

Section 2.4 Measures of Variation

► Example 5 (pg. 84)	Using Technology to Find the Standard Deviation

1. Open worksheet "Rent Rates" in the Chapter 2 folder.

2. Click **Data** and select **Data Analysis**. Select **Descriptive Statistics** and click **OK**.

   If Data Analysis does not appear as a choice in the Data ribbon, you will need to load the Microsoft Excel Analysis ToolPak add-in. Follow the procedure in Section GS 8.1 before continuing.

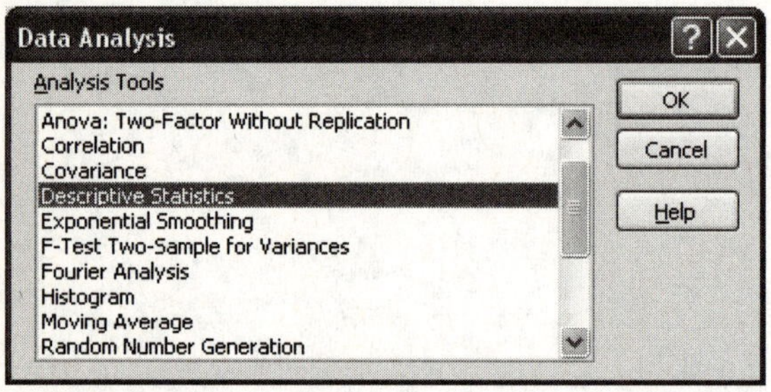

Data Analysis

Analysis Tools
- Anova: Two-Factor Without Replication
- Correlation
- Covariance
- Descriptive Statistics
- Exponential Smoothing
- F-Test Two-Sample for Variances
- Fourier Analysis
- Histogram
- Moving Average
- Random Number Generation

OK | Cancel | Help

3. Complete the Descriptive Statistics dialog box as shown below. Click **OK**.

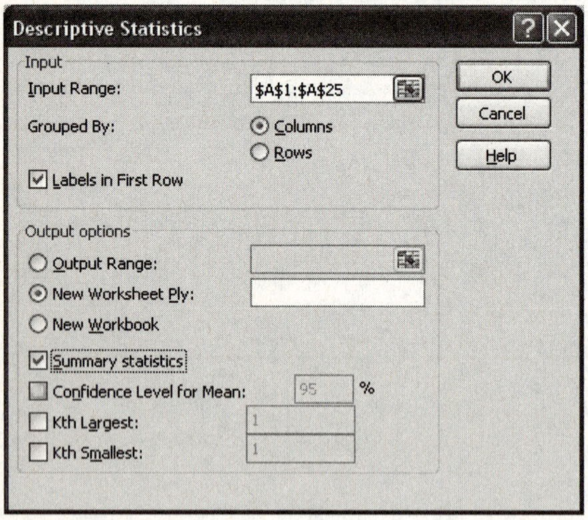

Descriptive Statistics

Input
Input Range: $A$1:$A$25
Grouped By: ⦿ Columns ○ Rows
☑ Labels in First Row

Output options
○ Output Range:
⦿ New Worksheet Ply:
○ New Workbook
☑ Summary statistics
☐ Confidence Level for Mean: 95 %
☐ Kth Largest: 1
☐ Kth Smallest: 1

OK | Cancel | Help

You will want to make Column A of the output wider so that you can read the labels. Your Descriptive Statistics output should be similar to the output shown at the top of the next page. The mean is 33.73 and the standard deviation is 5.09.

	A	B
1	*Office rental rates*	
2		
3	Mean	33.72917
4	Standard Error	1.038864
5	Median	35.375
6	Mode	37
7	Standard Deviation	5.089373
8	Sample Variance	25.90172
9	Kurtosis	-0.74282
10	Skewness	-0.70345
11	Range	16.75
12	Minimum	23.75
13	Maximum	40.5
14	Sum	809.5
15	Count	24

◀

Section 2.5 Measures of Position

▶ Example 2 (pg. 101)   Using Technology to Find Quartiles

1. Open worksheet "Tuition" in the Chapter 2 folder. Key in labels for the quartiles as shown below. Then click in cell **D2** to place the first quartile there.

	A	B	C	D
1	Tuition costs (thousands of dollars)		Quartile	
2	23		1	
3	25		2	
4	30		3	

2. You will be using the QUARTILE function to obtain the first, second, and third quartiles for the tuition data. At the top of the screen, click **Formulas** and select **Insert Function**.

3. Select the **Statistical** category. Select the **QUARTILE** function. Click **OK**.

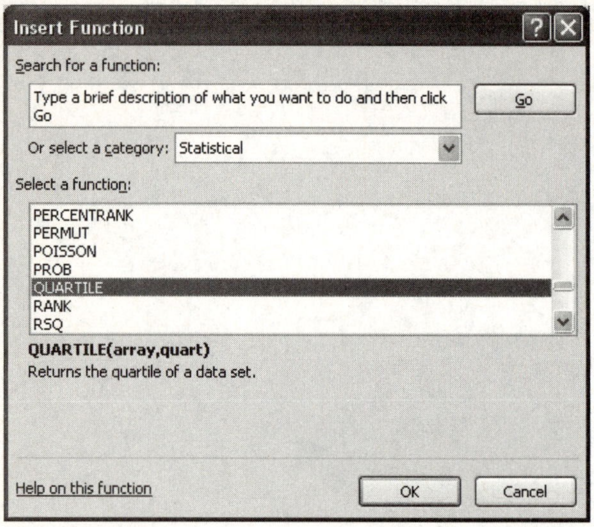

Insert Function

Search for a function:

| Type a brief description of what you want to do and then click Go | Go |

Or select a category: Statistical

Select a function:

PERCENTRANK
PERMUT
POISSON
PROB
QUARTILE
RANK
RSQ

QUARTILE(array,quart)
Returns the quartile of a data set.

Help on this function   OK   Cancel

4.   Complete the QUARTILE dialog box as shown below. Click **OK**.

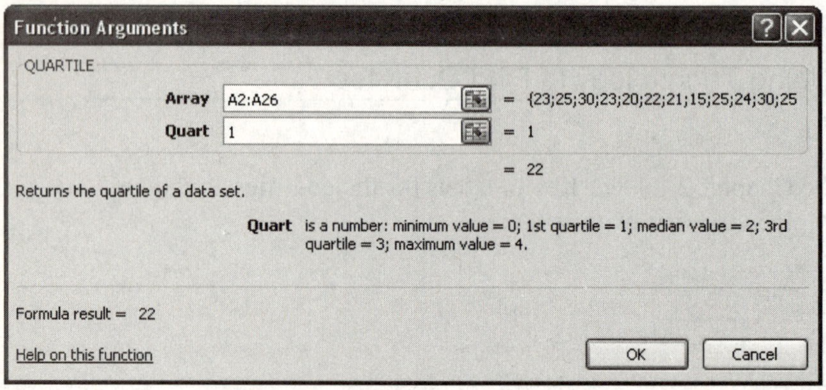

Function Arguments

QUARTILE

Array   A2:A26   = {23;25;30;23;20;22;21;15;25;24;30;25

Quart   1   = 1

= 22

Returns the quartile of a data set.

Quart is a number: minimum value = 0; 1st quartile = 1; median value = 2; 3rd quartile = 3; maximum value = 4.

Formula result =   22

Help on this function   OK   Cancel

5.   Click in cell **D3** to place the second quartile there.

6.   At the top of the screen, click **Formulas** and select **Insert Function**.

7.   Select the **Statistical** category and the **QUARTILE** function. Click **OK**.

8.   Complete the QUARTILE dialog box as shown at the top of the next page. Click **OK**.

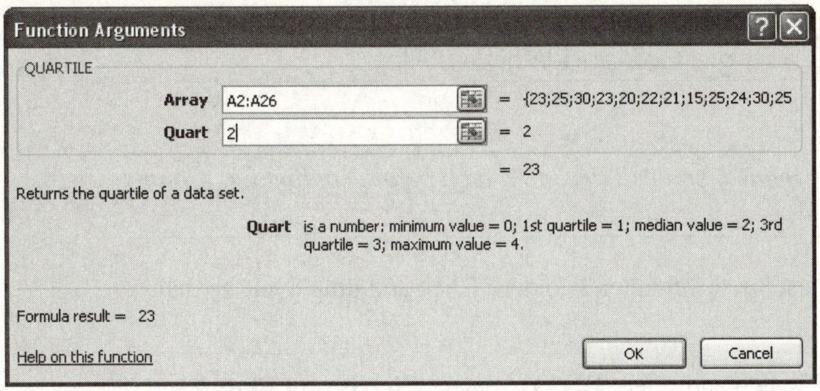

Function Arguments   [?][X]

QUARTILE

    Array   A2:A26   [▦]   = {23;25;30;23;20;22;21;15;25;24;30;25

    Quart   2| [▦]   = 2

                       = 23

Returns the quartile of a data set.

               Quart is a number: minimum value = 0; 1st quartile = 1; median value = 2; 3rd
                       quartile = 3; maximum value = 4.

Formula result = 23

Help on this function   [OK] [Cancel]

9. Click in cell **D4** to place the third quartile there.

10. At the top of the screen, click **Formulas** and select **Insert Function**.

11. Select the **Statistical** category and the **QUARTILE** function. Click **OK**.

12. Complete the QUARTILE dialog box as shown below. Click **OK**.

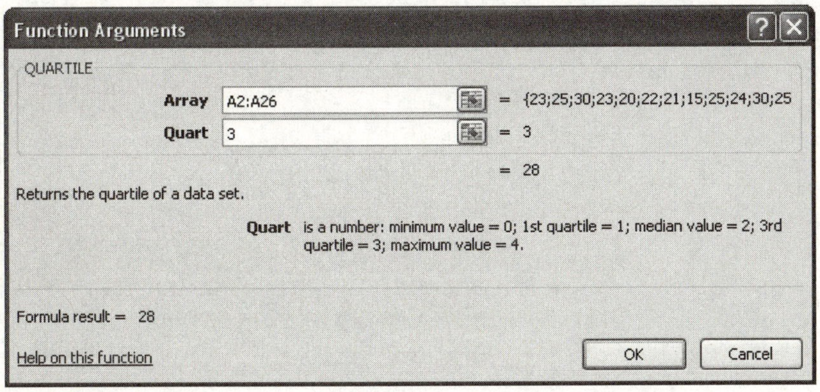

Function Arguments   [?][X]

QUARTILE

    Array   A2:A26   [▦]   = {23;25;30;23;20;22;21;15;25;24;30;25

    Quart   3   [▦]   = 3

                       = 28

Returns the quartile of a data set.

               Quart is a number: minimum value = 0; 1st quartile = 1; median value = 2; 3rd
                       quartile = 3; maximum value = 4.

Formula result = 28

Help on this function   [OK] [Cancel]

Your output should look similar to the output displayed below.

	A	B	C	D
1	Tuition costs (thousands of dollars)		Quartile	
2	23		1	22
3	25		2	23
4	30		3	28

▶ Example 4 (pg. 103) Drawing a Box-and-Whisker Plot

If the DDXL add-in has not been loaded, you will need to load it before continuing. Follow the instructions in Section GS 8.2.

1. Open worksheet "World's Richest" in the chapter 2 folder. Click and drag over the data range A1:A51.

2. Click **Add-Ins** at the top of the screen and select **DDXL**. Select **Charts and Plots**.

3. Click the down arrow under Function type and select **Boxplot**.

```
Charts and Plots Dialog

Function type:
** CLICK ME **              ▼

** CLICK ME **
3D Rotating Plot
Bar Chart
Boxplot
Boxplot by Groups
Color Scatterplot
Dotplot
Dotplot by Groups
Histogram
Normal Probability Plot
Pie Chart
Scatterplot
StackedDotplot
```

4. Select **Ages** in the Names and Columns window. Then click the arrow under Quantitative Variables on the left to select Ages for the box plot. Note that a checkmark has already been placed in the box next to **First row is variable names**. Click **OK**.

5. Scroll the Summary Statistics output below the box plot to see the interquartile range, the first quartile, and the third quartile.

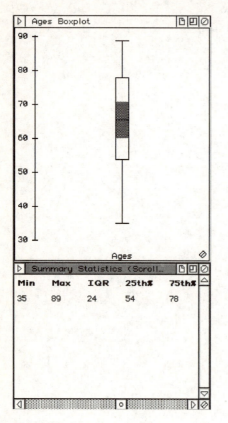

Technology

<table>
<tr><td>▶ Exercises 1 & 2   (pg. 121)</td><td>Finding the Sample Mean and the Sample Standard Deviation</td></tr>
</table>

1. Open worksheet "Tech2" in the Chapter 2 folder.

2. Click **Data** and select **Data Analysis**. Select **Descriptive Statistics** and click OK.

If Data Analysis does not appear as a choice in the Data ribbon, you will need to load the Microsoft Excel Analysis ToolPak add-in. Follow the procedure in Section GS 8.1 before continuing.

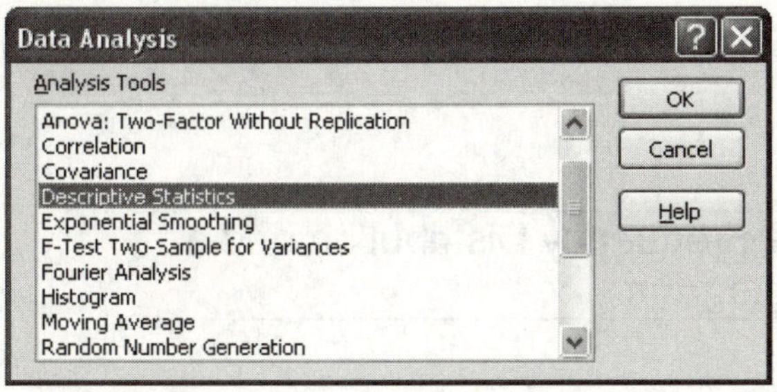

Data Analysis [?][X]

Analysis Tools

Anova: Two-Factor Without Replication
Correlation
Covariance
Descriptive Statistics
Exponential Smoothing
F-Test Two-Sample for Variances
Fourier Analysis
Histogram
Moving Average
Random Number Generation

[OK]
[Cancel]
[Help]

3. Complete the Descriptive Statistics dialog box as shown below. Click **OK**.

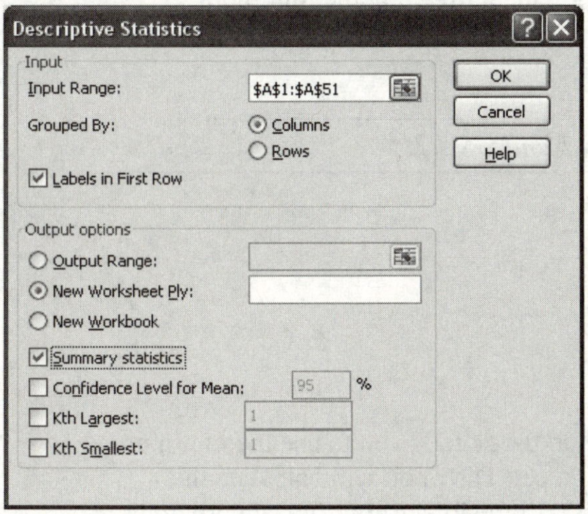

Descriptive Statistics [?][X]

Input

Input Range: $A$1:$A$51

Grouped By: ⦿ Columns
          ○ Rows

☑ Labels in First Row

Output options

○ Output Range:
⦿ New Worksheet Ply:
○ New Workbook
☑ Summary statistics
☐ Confidence Level for Mean: 95 %
☐ Kth Largest: 1
☐ Kth Smallest: 1

[OK]
[Cancel]
[Help]

You will want to adjust the width of column A of the output so that you can read the labels. Your Descriptive Statistics output should be similar to the output shown below. The sample mean is 2270.54 and the sample standard deviation is 653.1822.

	A	B
1	Monthly production (in pounds)	
2		
3	Mean	2270.54
4	Standard Error	92.37391
5	Median	2207
6	Mode	2207
7	Standard Deviation	653.1822
8	Sample Variance	426647
9	Kurtosis	0.567664
10	Skewness	0.549267
11	Range	3138
12	Minimum	1147
13	Maximum	4285
14	Sum	113527
15	Count	50

◀

▶ Exercises 3 & 4 (pg. 121) | Constructing a Frequency Distribution and a Frequency Histogram

---

1. Open worksheet "Tech2" in the Chapter 2 folder.

2. Sort the data in ascending order. In the sorted data set you can see that the minimum production is 1147 pounds. Scroll down to find the maximum production. The maximum production, displayed in cell A51, is 4285 pounds.

For instructions on how to sort, refer to Sections GS 5.1 and GS 5.2.

	A
1	Monthly production (in pounds)
2	1147
3	1230
50	3512
51	4285

3. Enter **Lower Limit** in cell D1 of the worksheet. The textbook tells you to use the minimum value as the lower limit of the first class. Enter **1147** in cell D2. You will calculate the remaining lower limits by adding the class width of 500 to the lower limit of each previous class. You will use a formula to do these computations in the Excel worksheet. Click in cell **D3** and key in **=D2+500** as shown at the top of the next page. Press [**Enter**].

	A	B	C	D
1	Monthly production (in pounds)			Lower Limit
2	1147			1147
3	1230			=D2+500

4. Click in cell **D3** (where 1647 now appears) and copy the contents of cell D3 to cells D4 through D9. Because the maximum milk production is 4285 pounds, you have calculated one more lower limit than is needed for the histogram. The value of 4647, however, will be used when calculating the upper limit of the last class.

	A	B	C	D
1	Monthly production (in pounds)			Lower Limit
2	1147			1147
3	1230			1647
4	1258			2147
5	1294			2647
6	1319			3147
7	1449			3647
8	1619			4147
9	1647			4647

5. Enter **Upper Limit** in cell E1. The upper limit is equal to one less than the lower limit of the next higher class. To do these calculations, you will enter a formula in the Excel worksheet. Click in cell E2 and enter the formula **=D3-1** as shown in the worksheet below. Press [**Enter**].

	A	B	C	D	E	F
1	Monthly production (in pounds)			Lower Limit	Upper Limit	
2	1147			1147	=D3-1	
3	1230			1647		

6. Copy the formula in E2 (where 1646 now appears) to cells E3 through E8. You will use these upper limits for the bins when you construct the histogram chart.

	A	B	C	D	E	F
1	Monthly production (in pounds)			Lower Limit	Upper Limit	
2	1147			1147	1646	
3	1230			1647	2146	
4	1258			2147	2646	
5	1294			2647	3146	
6	1319			3147	3646	
7	1449			3647	4146	
8	1619			4147	4646	

7. At the top of the screen, click **Data** and select **Data Analysis**. Select **Histogram** and click **OK**.

If Data Analysis does not appear as a choice in the Data ribbon, you will need to load the Microsoft Excel Analysis ToolPak add-in. Follow the procedure in Section GS 8.1 before continuing.

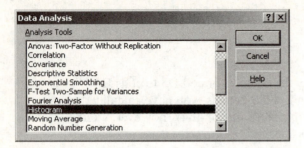

8. Complete the fields in the histogram dialog box as shown below. Click **OK**.

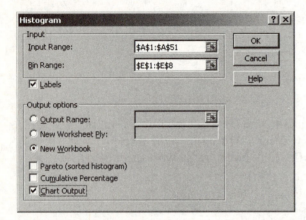

9. You will now follow steps to modify the histogram so that it is presented in a more informative manner. Begin by making the chart taller so that it is easier to read. To do this, first click within the figure near a border. Dotted handles appear. Click on the center handle at the bottom border of the figure and drag it down a few rows.

	A	B	C	D	E	F	G	H	I	J
1	Upper Limit	Frequency								
2	1646	7								
3	2146	15								
4	2646	13								
5	3146	11								
6	3646	3								
7	4146	0								
8	4646	1								
9	More	0								
10										
11										

10. To remove the "More" category from the X axis, first **right-click** on a vertical bar in the histogram and select **Select Data** from the shortcut menu that appears.

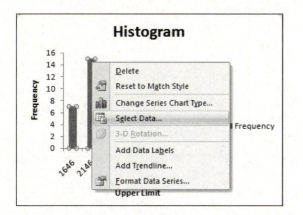

11. Click **Edit** under Horizontal (Category) Axis Labels.

12. "More" appears in cell A9 of the worksheet. To exclude the information in row 9 from the chart, change the 9 to 8 in the Axis label range window. Click **OK**. Click **OK** in the Select Data Source dialog box.

13. Next, change the chart title and the X-axis title. Click on "Histogram" and enter **Monthly Milk Production of 50 Holstein Dairy Cows**. Click on "Upper Limit" and enter **Pounds**.

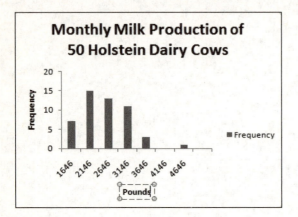

14. To remove the legend, click on the **Frequency** legend at the right and press [**Delete**].

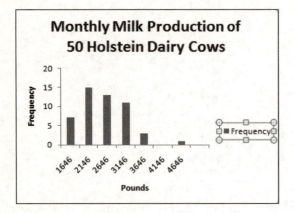

15. To add gridlines, **right-click** directly on a number in the Y-axis. Select I clicked on 20. Select **Add Major Gridlines**.

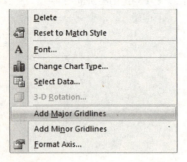

16. To remove the space between the vertical bars, **right-click** on one of the vertical bars and select **Format Data Series** from the shortcut menu that appears.

17. Move the Gap Width arrow all the way to the left so that 0% shows in the window. Click **Close**.

Your completed histogram should look similar to the one displayed at the top of the next page.

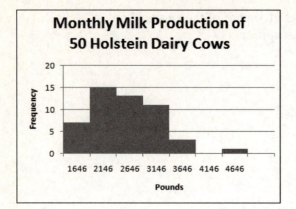

Probability

CHAPTER 3

Section 3.2 Conditional Probability and the Multiplication Rule

► Exercise 40 (pg. 155)	Simulating the "Birthday Problem"

Part d of Exercise 40 asks you to use a technology tool to simulate the "Birthday Problem." You will be generating 24 random numbers between 1 and 365.

If Data Analysis does not appear as a choice in the Data ribbon, you will need to load the Microsoft Excel Analysis ToolPak add-in. Follow the procedure in Section GS 8.1 before continuing.

---

1. Open a new Excel worksheet.

2. You will be entering the numbers 1 through 365 in the worksheet. To do this, you will fill column A with a number series. Begin by entering the numbers 1 and 2 in column A as shown below. You are starting a number series that will increase in increments of one.

	A
1	1
2	2

3. You will now complete the series all the way to 365. Begin by clicking in cell A1 and dragging down to cell A2 so that both cells are highlighted.

4. Move the mouse pointer in cell A2 to the bottom right corner of that cell so that the white plus sign turns into a black plus sign. This black plus sign is called the "fill handle." While the fill handle is displayed, hold down on the left mouse button and drag down column A until you reach cell A365. Release the mouse button. You should see the numbers from 1 to 365 in column A.

359	359
360	360
361	361
362	362
363	363
364	364
365	365

5. At the top of the screen, click **Data** and then select **Data Analysis** in the Analysis group. Select **Sampling** and click **OK**.

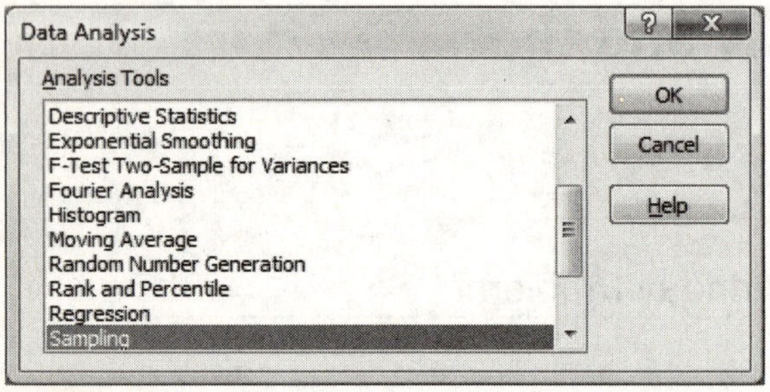

6. Complete the Sampling dialog box as shown below. Click **OK**.

You will be generating 10 different samples of 24 numbers. The first set of 24 numbers will be placed in column B, the second in column C, etc.

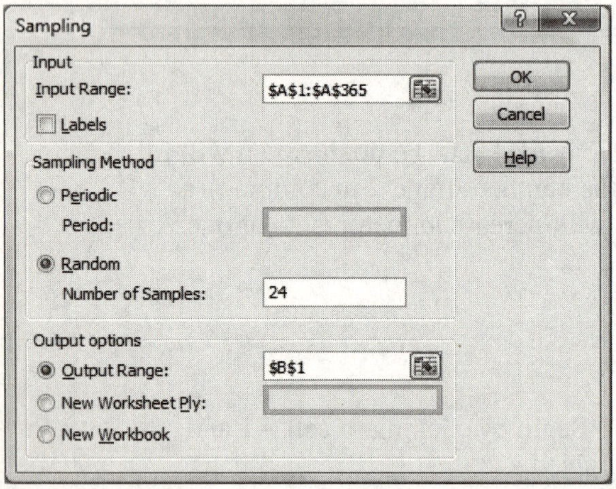

7. The 24 numbers, generated randomly with replacement, are displayed in column B. (Because the numbers were generated randomly, it is not likely that your output will be exactly the same.)

	A	B
1	1	87
2	2	113
3	3	276
4	4	54
5	5	19
6	6	360
7	7	277
8	8	290
9	9	350
10	10	315
11	11	331
12	12	262
13	13	359
14	14	195
15	15	132
16	16	298
17	17	266
18	18	204
19	19	269
20	20	48
21	21	194
22	22	124
23	23	273
24	24	241

If the DDXL add-in has not been loaded, you will need to load it before continuing. Follow the instructions in Section GS 8.2.

8. Next you will construct a frequency distribution to see if there are any repetitions. Be sure the range B1:B24 is highlighted. Click **Add-Ins** at the top of the screen and then click **DDXL**. Select **Tables**.

9. Click the down arrow under Function type and select **Frequency Table**. Note that there should not be a checkmark in the box next to First row is variable names.

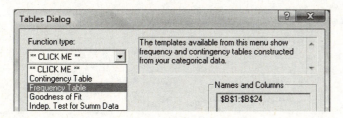

10.   Select the range **$B$1:$B$24** in the Names and Columns window by clicking on it. Then
      click the arrow under Categorical Variable on the left to select the data in **$B$1:$B$24** for
      the frequency table. Click **OK**.

```
Tables Dialog                                                    ?   x

 Function type:              This command displays a frequency table. You
                             need a column that holds category names.
 Frequency Table        ▼
                             Select the variable you want to display and drag it ▾

 ┌ Categorical Variable ──┐        ┌ Names and Columns ──┐
 │ $B$1:$B$24          │    ◄:┤    │ $B$1:$B$24          │
 │                        │          │                      │
 │ =!$B$1:$B$24  │  🖉  🗑  ▶        │                      │
                                    │                      │
                                    │                      │
                                    │                      │
                                    │                      │
                                    │                      │
                                    │                      │
                                    └──────────────────────┘
                                    ☐ First row is variable names

                                        Info          Help

                                        Cancel         OK
```

11.   You can see that there are no repetitions in this sample because all values have a count of 1.

```
▷ │ Summary of $B$1:$B$24           🗎 🔲
          Total Cases      24
 Number of Categories     24


▷ │ $B$1:$B$24 Frequency Table      🗎 🔲
 Group    Count    %
 19        1       4.167
 48        1       4.167
 54        1       4.167
 87        1       4.167
 113       1       4.167
 124       1       4.167
 132       1       4.167
 194       1       4.167
 195       1       4.167
 204       1       4.167
 241       1       4.167
 262       1       4.167
 266       1       4.167
 269       1       4.167
 273       1       4.167
 276       1       4.167
 277       1       4.167
 290       1       4.167
 298       1       4.167
```

12.   Close the Data Desk Viewer and go back to the sheet containing the numbers 1 through 365
      by clicking on the **Sheet1** tab near the bottom of the screen. Now you will generate the

second sample. Click **Data** at the top of the screen and select **Data Analysis**. Select **Sampling** and click **OK**. Complete the Sampling dialog box as shown below. The second set of randomly generated numbers will be placed in column C. Click **OK**.

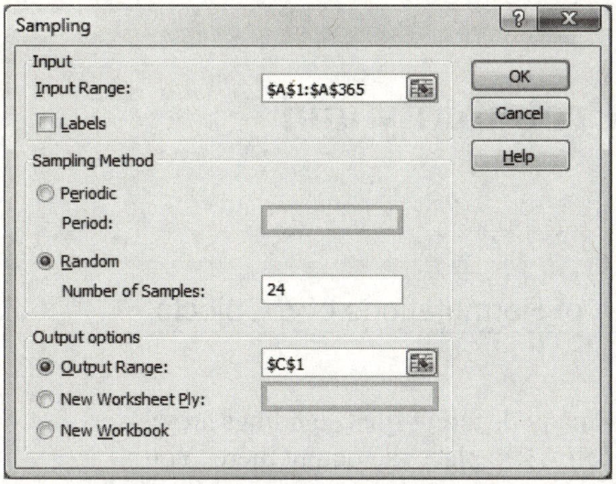

Sampling		? X
Input		
Input Range:	$A$1:$A$365	OK
☐ Labels		Cancel
Sampling Method		Help
○ Periodic		
Period:		
⦿ Random		
Number of Samples:	24	
Output options		
⦿ Output Range:	$C$1	
○ New Worksheet Ply:		
○ New Workbook		

13. Construct a frequency distribution to see if there are any repetitions in this second sample. Be sure the range C1:C24 is highlighted. Click **Add-Ins** at the top of the screen and then click **DDXL**. Select **Tables**.

14. Under Function type, select the **Frequency Table**. Click on the range **$C$1:$C$24** in the Names and Column window. Then click the arrow under Category Variable on the left to select the data in **$C$1:$C$24** for the frequency table. Click **OK**.

▷	Summary of $C$1:$C$24		🗔 🗗
	Total Cases	24	
	Number of Categories	24	

▷	$C$1:$C$24 Frequency Table		🗔 🗗

Group	Count	%
25	1	4.167
30	1	4.167
61	1	4.167
80	1	4.167
95	1	4.167
102	1	4.167
107	1	4.167
125	1	4.167
135	1	4.167
144	1	4.167
159	1	4.167
179	1	4.167
192	1	4.167
196	1	4.167
197	1	4.167
217	1	4.167
229	1	4.167
241	1	4.167
261	1	4.167

This second sample also has no repetitions.

15. Repeat the appropriate steps until you have generated 10 different samples of 24 numbers.

◀

Section 3.4 Additional Topics in Probability and Counting

▶ Example 1 (pg. 168)   Finding the Number of Permutations of *n* Objects

1. In Try It Yourself 1, you are asked to determine how many different final standings are possible. Open a new Excel worksheet and click in cell **A1** to place the output there. You will be using the PERMUT function to calculate the number of permutations.

2. At the top of the screen, click **Formulas** and select **Insert Function** in the Function Library group.

3. Select the **Statistical** category. Select the **PERMUT** function. Click **OK**.

4. Complete the PERMUT dialog box as shown at the top of the next page. There are eight women's hockey teams. The number of different final standings is equal to 8! Click **OK**.

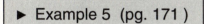

Function Arguments

PERMUT

Number	8	= 8
Number_chosen	8	= 8

= 40320

Returns the number of permutations for a given number of objects that can be selected from the total objects.

Number_chosen   is the number of objects in each permutation.

Formula result =   40320

Help on this function   OK   Cancel

There are 40,320 possible different final standings.

◄

► Example 5 (pg. 171)	Finding the Number of Combinations

1. You are asked to calculate how many ways the manager can form a three-person advisory committee from 20 employees. You will be using the COMBIN function to calculate the number of combinations. Open a new Excel worksheet and click in cell **A1** to place the output there.

2. At the top of the screen, click **Formulas** and select **Insert Function** in the Function Library group.

3.   Select the **Math & Trig** category. Select the **COMBIN** function. Click **OK**.

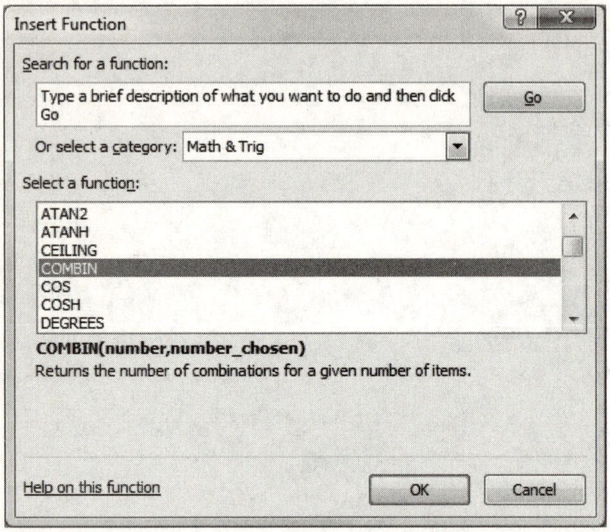

Insert Function		? X
Search for a function:		
Type a brief description of what you want to do and then click Go		Go
Or select a category: Math & Trig		▾
Select a function:		
ATAN2 ATANH CEILING **COMBIN** COS COSH DEGREES		
COMBIN(number,number_chosen) Returns the number of combinations for a given number of items.		
Help on this function	OK	Cancel

4.   Complete the COMBIN dialog box as shown below. Click **OK**.

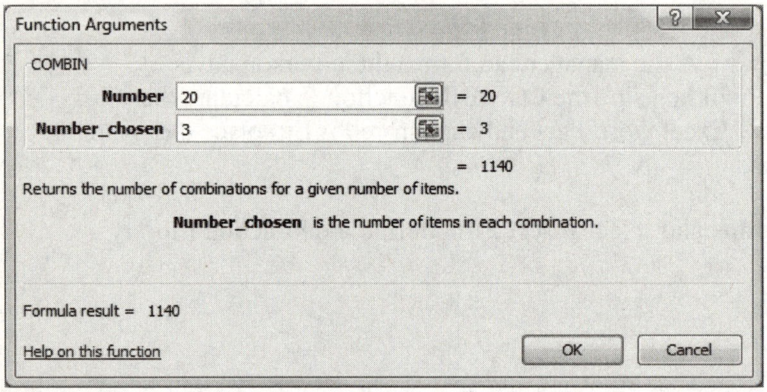

Function Arguments		? X
COMBIN		
Number	20	= 20
Number_chosen	3	= 3
		= 1140
Returns the number of combinations for a given number of items.		
	Number_chosen is the number of items in each combination.	
Formula result = 1140		
Help on this function	OK	Cancel

There are 1,140 different combinations.

◀

Technology

| ► Exercise 3 (pg. 187) | Selecting Randomly a Number from 1 to 11 |

If Data Analysis does not appear as a choice in the Data ribbon, you will need to load the Microsoft Excel Analysis ToolPak add-in. Follow the procedure in Section GS 8.1 before continuing.

You will be given the steps to follow to complete Exercise 3 and Exercise 3b.

1. You will first select randomly a number between 1 and 11. Open a new Excel worksheet. Enter the numbers 1 through 11 in cells A1 through A11 as shown below.

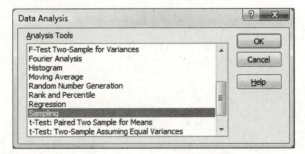

2. At the top of the screen, click **Data** and select **Data Analysis** in the Analysis group. Select **Sampling**, and click **OK**.

3. Complete the Sampling dialog box as shown below. Click **OK**.

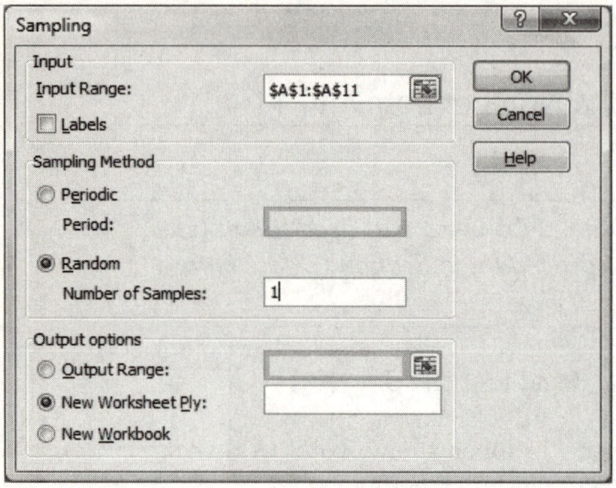

4. The output is displayed in a new sheet. The number 8 was selected. Because the number was selected randomly, your output may not be the same. You will now select randomly 100 integers from 1 to 11. Click the **Sheet1** tab near the bottom of the screen to go back to the worksheet displaying the numbers from 1 to 11.

5. At the top of the screen, click **Data** and select **Data Analysis** in the Analysis group. Select **Sampling**, and click **OK**.

6. Complete the Sampling dialog box as shown below. Click **OK**.

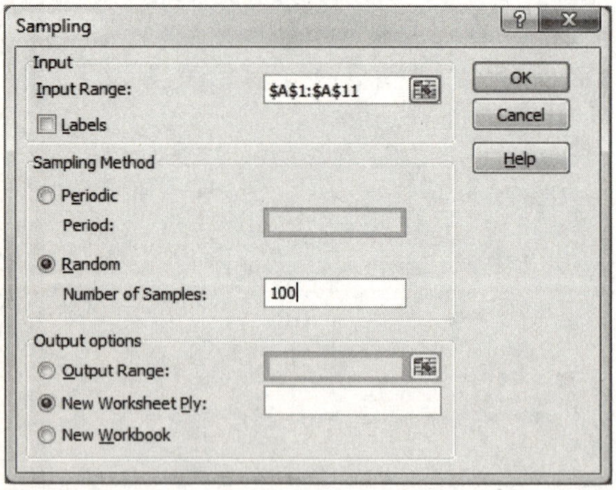

If the DDXL add-in has not been loaded, you will need to load it before continuing. Follow the instructions in Section GS 8.2.

7. The output is displayed in a new sheet. You will need to scroll down to see the entire listing of the 100 randomly generated integers. You will now tally the results. Be sure the range A1:A100 is highlighted. Then click **Add-Ins** at the top of the screen and click **DDXL**. Select **Tables**.

	A
1	11
2	1
3	10
4	4
5	3
6	8

8. Click the down arrow under Function type and select **Frequency Table**.

9. Remove the checkmark next to First row is variable names at the bottom of the dialog box. Select the range **$A$1:$A$100** in the Names and Columns window by clicking on it. Then click the arrow under Categorical Variable on the left to select the data in **$A$1:$A$100** for the frequency table. Click **OK**.

The output for Exercise 3b is displayed at the top of the next page. Your output should have the same format. Because the numbers were generated randomly, however, it is not likely that your numbers will be exactly the same.

```
┌────────────────────────────────────┐
│ ▷ Summary of $A$1:$A$100      ▣▣  │
│          Total Cases    100        │
│ Number of Categories     11        │
├────────────────────────────────────┤
│ ▷ $A$1:$A$100 Frequency Table  ▣▣ │
│ Group    Count    %                │
│ 1          8       8               │
│ 2          9       9               │
│ 3         13      13               │
│ 4         10      10               │
│ 5          5       5               │
│ 6          9       9               │
│ 7         10      10               │
│ 8          6       6               │
│ 9          6       6               │
│ 10        11      11               │
│ 11        13      13               │
│                                    │
└────────────────────────────────────┘
```

◀

► Exercise 5 (pg. 187)   Composing Mozart Variations with Dice

If Data Analysis does not appear as a choice in the Data ribbon, you will need to load the Microsoft Excel Analysis ToolPak add-in. Follow the procedure in Section GS 8.1 before continuing.

You will first be given the steps to follow to complete Exercise 5 and then the steps to follow to complete Exercise 5b. In Exercise 5, you will select randomly two integers from 1, 2, 3, 4, 5, and 6. This simulates tossing two six-sided dice one time. You will then sum the two integers and subtract 1 from the sum. This is Mozart's procedure (described at the top of page 187) for selecting a musical phrase from 11 different choices.

1. To select randomly two integers between 1 and 6, begin by opening a new Excel worksheet and entering the numbers 1 through 6 in column A as shown below.

	A
1	1
2	2
3	3
4	4
5	5
6	6

2. At the top of the screen, click **Data** and select **Data Analysis** in the Analysis group. Select **Sampling**, and click **OK**.

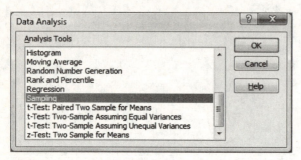

3. Complete the Sampling dialog box as shown below. Click **OK**.

4. The output is displayed in a new worksheet. The numbers 1 and 3 were selected. Mozart's instructions are to sum the numbers and subtract 1. The result is 3. Therefore, musical phrase number 3 would be selected for the first bar of Mozart's minuet.

	A
1	1
2	3

5. Exercise 5b asks you to simulate, 100 times, Mozart's procedure of tossing two die, finding the sum, and subtracting 1. You are also asked to tally the results. Begin by entering the numbers 1 through 11 in a new worksheet as shown below.

	A
1	1
2	2
3	3
4	4
5	5
6	6
7	7
8	8
9	9
10	10
11	11

6. At the top of the screen, click **Data** and select **Data Analysis** in the Analysis group. Select **Sampling**, and click **OK**.

7. Complete the Sampling dialog box as shown below. Click **OK**.

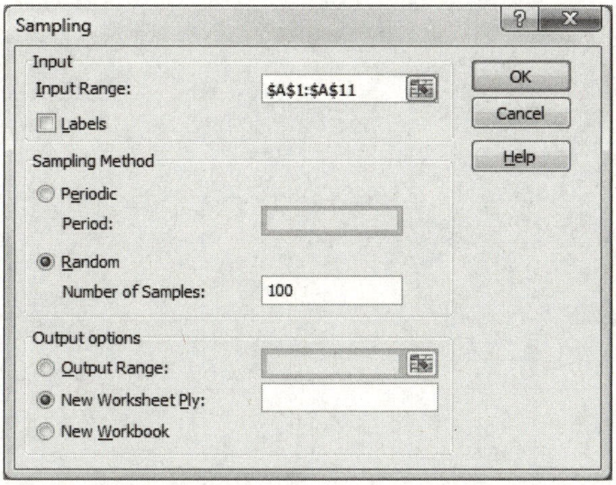

If the DDXL add-in has not been loaded, you will need to load it before continuing. Follow the instructions in Section GS 8.2.

---

8.   The output is displayed in a new sheet. You need to scroll down to see the entire listing of the 100 randomly generated integers. You will now tally the results. Be sure the range A1:A100 is highlighted. Then click **Add-Ins** at the top of the screen and click **DDXL**. Select **Tables**.

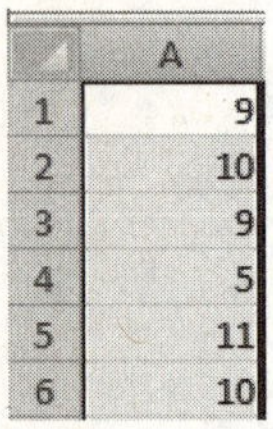

	A
1	9
2	10
3	9
4	5
5	11
6	10

9.   Click the down arrow under Function type and select **Frequency Table**.

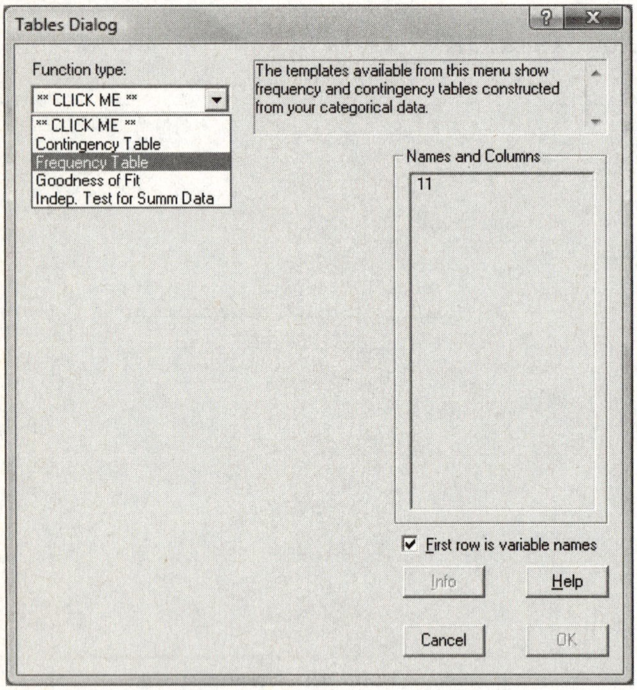

10. Remove the checkmark next to First row is variable names at the bottom of the dialog box. Select the range **$A$1:$A$100** in the Names and Columns window by clicking on it. Then click the arrow under Categorical Variable on the left to select the data in that range for the frequency table. Click **OK**.

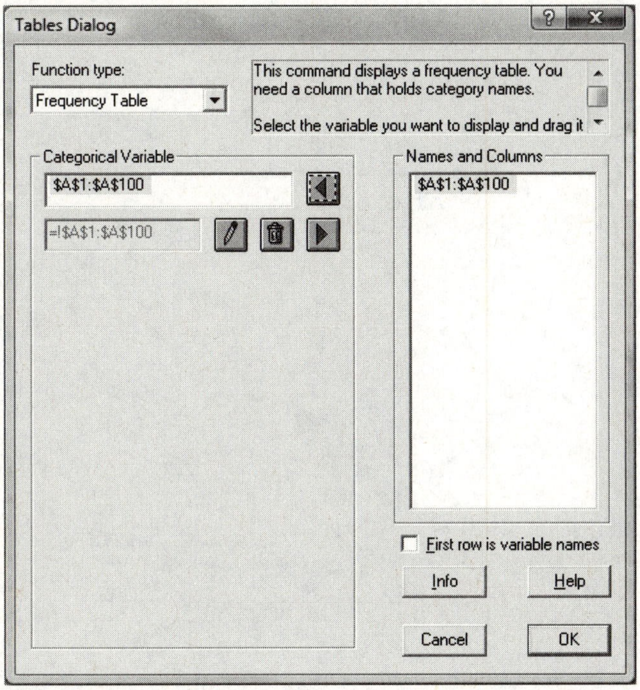

Tables Dialog	? ☐ ✕
Function type:	This command displays a frequency table. You need a column that holds category names.
Frequency Table ▼	Select the variable you want to display and drag it ▼

Categorical Variable

$A$1:$A$100

=!$A$1:$A$100   ✏️ 🗑️ ▶️

Names and Columns

$A$1:$A$100

☐ First row is variable names

Info   Help

Cancel   OK

The output for Exercise 5b is displayed at the top of the next page. Your output should have the same format. Because the numbers were generated randomly, however, it is not likely that your numbers will be exactly the same.

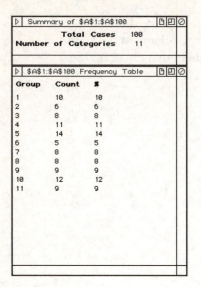

Group	Count	%
1	10	10
2	6	6
3	8	8
4	11	11
5	14	14
6	5	5
7	8	8
8	8	8
9	9	9
10	12	12
11	9	9

Discrete Probability Distributions

Section 4.2 Binomial Distributions

> ▶ Example 4 (pg. 206)   Finding a Binomial Probability Using Technology

1. In Try It Yourself 4, you are asked to find the probability that exactly 178 people will use more than one topping on their hot dogs. Open a new Excel worksheet and click in cell **A1** to place the output there.

2. Click **Formulas** at the top of the screen and select **Insert Function**.

3. Select the **Statistical** category. Select the **BINOMDIST** function. Click **OK**.

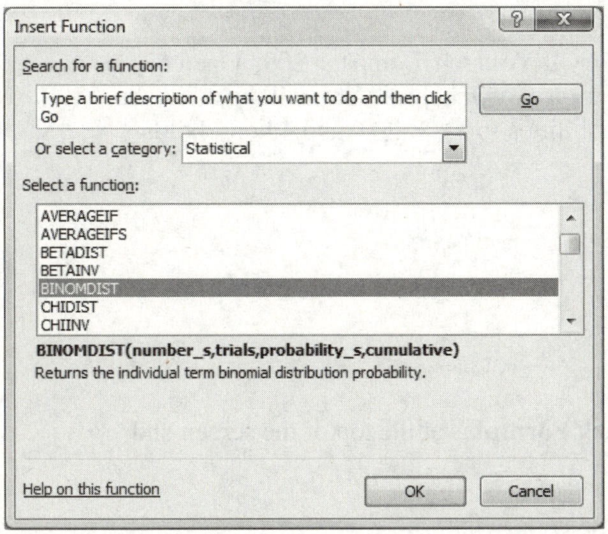

Insert Function

Search for a function:

Type a brief description of what you want to do and then click Go | Go |

Or select a category: Statistical

Select a function:

AVERAGEIF
AVERAGEIFS
BETADIST
BETAINV
BINOMDIST
CHIDIST
CHIINV

BINOMDIST(number_s,trials,probability_s,cumulative)
Returns the individual term binomial distribution probability.

Help on this function | OK | Cancel |

4. Complete the BINOMDIST dialog box as shown below. Click **OK**.

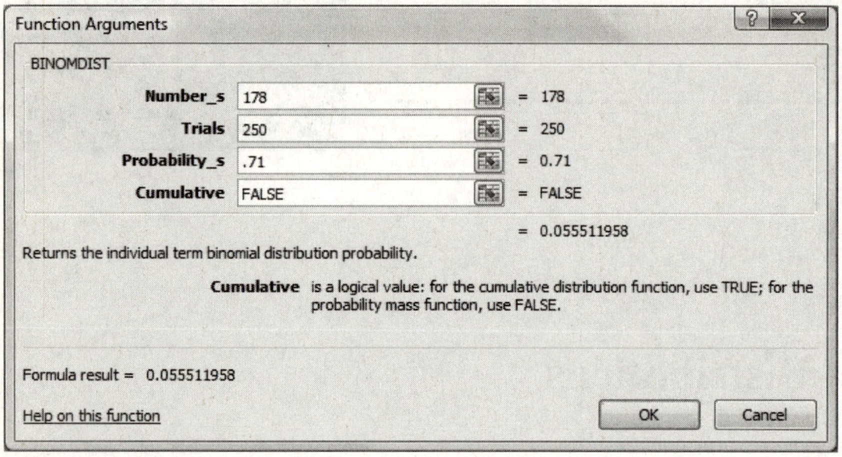

Function Arguments		? ⬛ ✖
BINOMDIST		
Number_s	178 ▦	= 178
Trials	250 ▦	= 250
Probability_s	.71 ▦	= 0.71
Cumulative	FALSE ▦	= FALSE
		= 0.055511958

Returns the individual term binomial distribution probability.

       Cumulative is a logical value: for the cumulative distribution function, use TRUE; for the probability mass function, use FALSE.

Formula result = 0.055511958

Help on this function   [OK]   [Cancel]

The BINOMDIST function returns a result of 0.055512.

	A
1	0.055512

◀

▶ Example 7 (pg. 209)	Graphing a Binomial Distribution

1. You will be graphing a binomial distribution for Try It Yourself 7 on page 209. Open a new, blank Excel worksheet and enter the information shown below. You will be using the BINOMDIST function to calculate binomial probabilities for 0, 1, 2, 3, and 4 households.

	A	B	C
1	Households	Relative frequency	
2	0		
3	1		
4	2		
5	3		
6	4		

2. Click in cell **B2** below "Relative frequency." Click **Formulas** at the top of the screen and select **Insert Function**.

3. Under Function category, select **Statistical**. Under Function name, select **BINOMDIST**. Click **OK**.

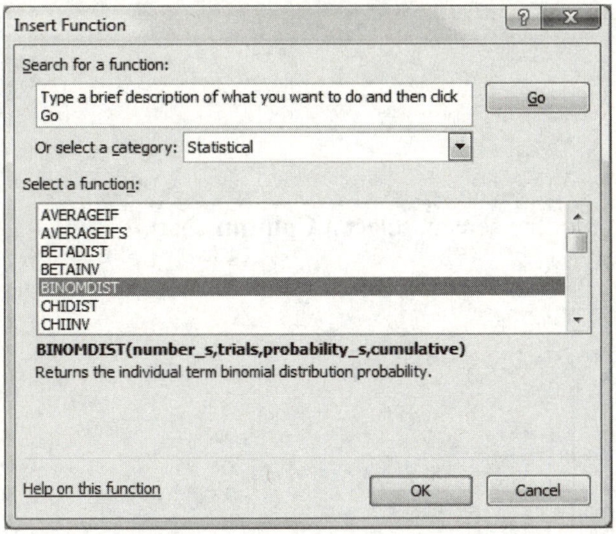

4. Complete the BINOMDIST dialog box as shown below. Click **OK**.

You are entering a relative cell address without dollar signs (i.e., A2) in the Number_s field because you will be copying the contents of cell B2 to cells B3 through B6. You want the column A cell address to change from A2 to A3, A4, …, A6 when the formula is copied from cell B2 to cells B3 through B6.

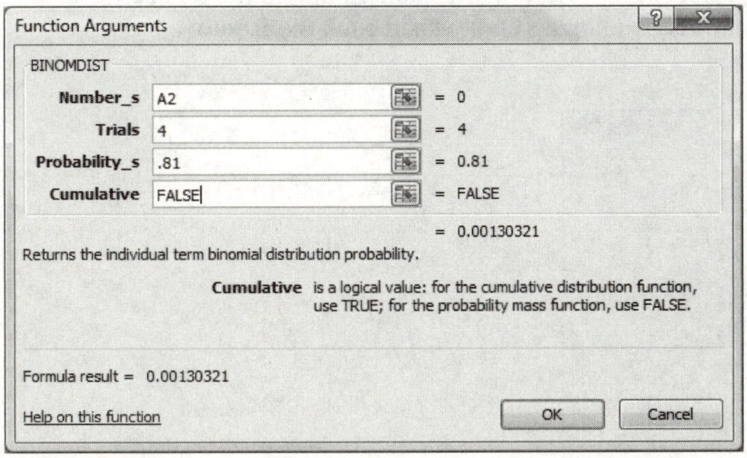

5.   Copy the contents of cell B2 to cells B3 through B8.

	A	B	C
1	Households	Relative frequency	
2	0	0.00130321	
3	1	0.02222316	
4	2	0.14211126	
5	3	0.40389516	
6	4	0.43046721	

6.   Highlight cells **A1:B6**. Then click **Insert** at the top of the screen, select a **Column** chart, and select the leftmost diagram in the 2-D Column row.

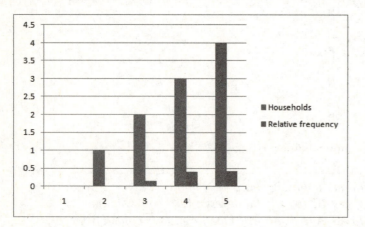

7.   The chart needs to be modified so that only the relative frequency bars are displayed. Click **Design** near the top of the screen and select **Select Data** in the Data group. In the Select Data Source dialog box, select **Households** by clicking on it. Then click the **Remove** button. Click **OK**.

8. Click on the **Relative Frequency** title at the top of the chart and change it to **Owning a Video Game Console**.

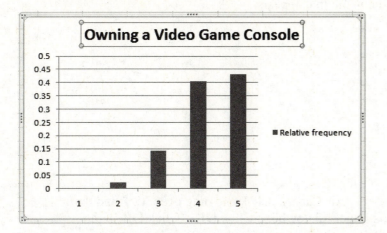

9. At the top of the screen, in the Chart Layouts group, select **Layout 8** so that you can add a Y-axis title and an X-axis title.   Change the Y-axis title to **Relative Frequency** and the X-axis title to **Households**.

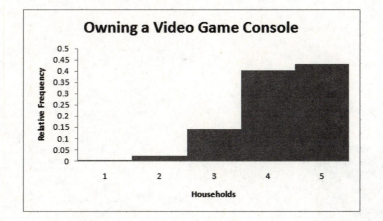

10. The X-axis number scale should be 0 to 4. These are the numbers in column A of the worksheet. To make this change, **right-click** on any number in the X-axis in the chart. I clicked on 3. Click on **Select Data** in the menu that appears.

	<u>D</u>elete
	Reset to M<u>a</u>tch Style
A	<u>F</u>ont...
	Change Chart T<u>y</u>pe...
	<u>S</u>elect Data...
	3-D <u>R</u>otation...
	Add <u>M</u>ajor Gridlines
	Add Mi<u>n</u>or Gridlines
	<u>F</u>ormat Axis...

11. Click on **Edit** below Horizontal (Category) Axis Labels. Click and drag over A2:A8 in the worksheet to enter that range in the Axis Labels dialog box. Click **OK**. Click **OK** in the Select Data Source dialog box.

Axis Labels	? X
<u>A</u>xis label range:	
=Sheet1!$A$2:$A$6	= 0, 1, 2, 3, 4
OK	Cancel

The completed chart is shown below.

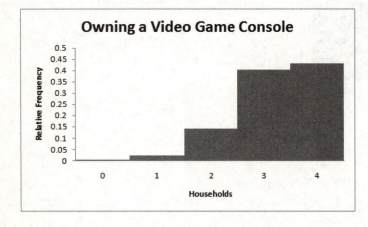

Section 4.3 More Discrete Probability Distributions

| ► Example 2 (pg. 219) | Using the Poisson Distribution |

You will be using the POISSON function to find the probability that 4 accidents will occur at a certain intersection when the mean number of accidents per month at this intersection is 3.

1. Open a new Excel worksheet and click in cell **A1** to place the output there.

2. At the top of the screen, click **Formulas** and select **Insert Function**.

3. Select the **Statistical** category. Select the **POISSON** function. Click **OK**.

Insert Function

Search for a function:

Type a brief description of what you want to do and then click Go   Go

Or select a category: Statistical

Select a function:

PERCENTILE
PERCENTRANK
PERMUT
POISSON
PROB
QUARTILE
RANK

POISSON(x,mean,cumulative)
Returns the Poisson distribution.

Help on this function   OK   Cancel

4. Complete the POISSON dialog box as shown at the top of the next page. Click **OK**.

The POISSON function returns a result of 0.168031.

◄

Technology

► Exercise 1 (pg. 233)	Creating a Poisson Distribution with μ = 4 for x = 0 to 20

1. Open a new Excel worksheet and enter the numbers 0 through 20 in column A as shown at the top of the next page. Then click in cell **B1** of the worksheet to place the output from the POISSON function there.

	A	B
1	0	
2	1	
3	2	
4	3	
5	4	
6	5	
7	6	
8	7	
9	8	
10	9	
11	10	
12	11	
13	12	
14	13	
15	14	
16	15	
17	16	
18	17	
19	18	
20	19	
21	20	

2.   At the top of the screen, click **Formulas** and select **Insert Function**.

3.   Select the **Statistical** category. Select the **POISSON** function. Click **OK**.

4. Complete the POISSON dialog box as shown below. Click **OK**.

You are entering a relative cell address without dollar signs (i.e., A1) in the X field because you will be copying the contents of cell B1 to cells B2 through B21. You want the column A cell address to change from A1 to A2, A3, A4, …, A21 when the formula is copied from cell B1 to cells B2 through B21.

Function Arguments		? X
POISSON		
X	A1	= 0
Mean	4	= 4
Cumulative	FALSE	= FALSE
		= 0.018315639

Returns the Poisson distribution.

Cumulative is a logical value: for the cumulative Poisson probability, use TRUE; for the Poisson probability mass function, use FALSE.

Formula result = 0.018315639

Help on this function   [OK]   [Cancel]

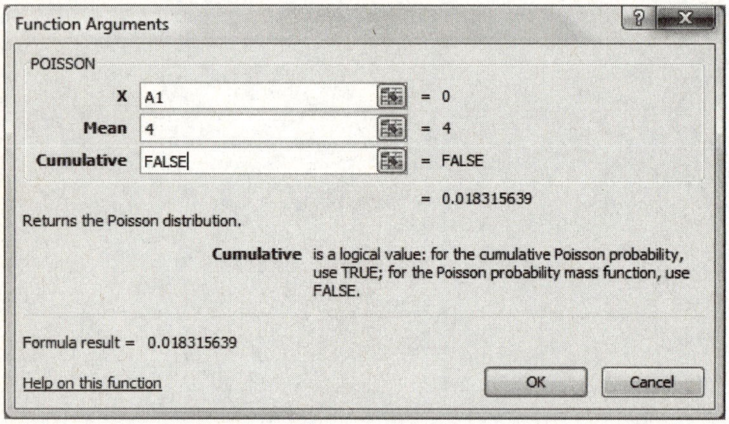

5. Copy the contents of cell B1 to cells B2 through B21.

	A	B
1	0	0.018316
2	1	0.073263
3	2	0.146525
4	3	0.195367
5	4	0.195367
6	5	0.156293
7	6	0.104196
8	7	0.05954
9	8	0.02977
10	9	0.013231
11	10	0.005292
12	11	0.001925
13	12	0.000642
14	13	0.000197
15	14	5.64E-05
16	15	1.5E-05
17	16	3.76E-06
18	17	8.85E-07
19	18	1.97E-07
20	19	4.14E-08
21	20	8.28E-09

6. Construct a histogram of the Poisson distribution. First highlight cells **A1:B21**. Then, at the top of the screen, click **Insert** and select a **Column** chart.

7.   Select the leftmost diagram in the 2-D row.

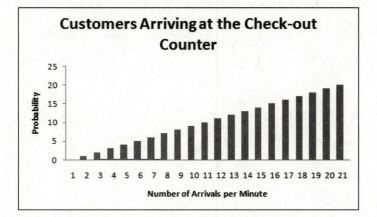

8.   In the Chart Layouts section near the top of the screen, select **Layout 8** so that you can add a chart title, Y-axis title, and X-axis title. Click on each title to change it. Change the chart title to **Customers Arriving at the Check-out Counter,** change the Y-axis title to **Probability**, and change the X-axis title to **Number of Arrivals per Minute**.

9.   Next you will modify the chart so that the correct values are displayed. **Right-click** on one of the vertical bars. Then click on **Select Data** in the shortcut menu that appears.

10.   Click on **Series1.** Then click the **Remove** button.

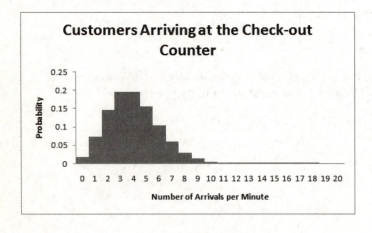

11.   The X-axis number scale should be 0 to 20. These are the numbers in column A of the worksheet. To make this change, click on **Edit** below Horizontal (Category) Axis Labels on the right side of the dialog box. Click and drag over A1:A21 in the worksheet to enter that range in the Axis Labels dialog box. Click **OK**.   Click **OK** in the Select Data Source dialog box.

The completed chart is shown below.

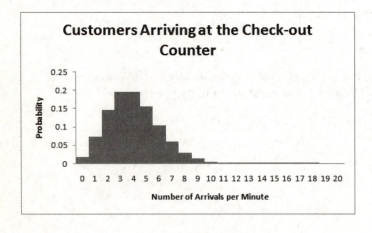

▶ Exercise 3 (pg. 233)	Generating a List of 20 Random Numbers with a Poisson Distribution for μ = 4

1. Open a new, blank Excel worksheet. At the top of the screen, click **Data** and select **Data Analysis**. Select **Random Number Generation** and click **OK**.

If Data Analysis does not appear as a choice in the Data ribbon, you will need to load the Microsoft Excel Analysis ToolPak add-in. Follow the procedure in Section GS 8.1 before continuing.

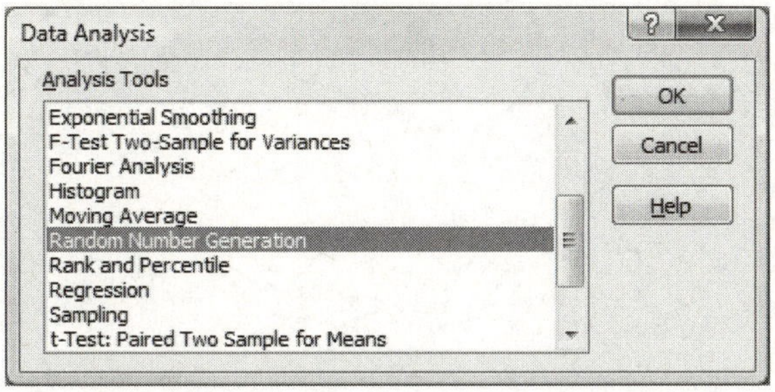

---

2. Complete the Random Number Generation dialog box as shown below. The **Number of Variables** indicates the number of columns of values that you want in the output table. **Number of Random Numbers** to be generated is 20. Select the **Poisson** distribution. **Lambda** is the expected value (μ), which is equal to 4. The output will be placed in the current worksheet with A1 as the left topmost cell. Click **OK**.

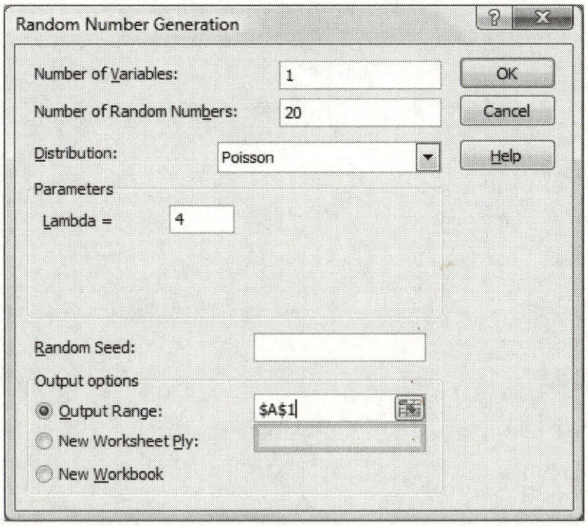

The 20 numbers generated for this example are shown below. Because the numbers were generated randomly, it is not likely that your numbers will be exactly the same.

	A
1	5
2	3
3	2
4	4
5	6
6	4
7	3
8	2
9	2
10	0
11	3
12	2
13	4
14	6
15	2
16	8
17	6
18	3
19	3
20	3

◀

Normal Probability Distributions

Section 5.1 Introduction to Normal Distributions and the Standard Normal Distribution

▶ Example 4 (pg. 242)	Finding Area Under the Standard Normal Curve

1. You are asked to find the area under the standard normal curve to the left of $z = -0.99$. Open a new Excel worksheet and click in cell **A1** to place the output there.

2. At the top of the screen, click **Formulas** and select **Insert Function**.

3. Select the **Statistical** category. Select the **NORMSDIST** function as shown at the top of the next page. Click **OK**.

Be sure to select NORMSDIST and not NORMDIST.

Insert Function ? X

Search for a function:

| Type a brief description of what you want to do and then click Go | Go |

Or select a category: Statistical ▾

Select a function:

```
MODE
NEGBINOMDIST
NORMDIST
NORMINV
NORMSDIST
NORMSINV
PEARSON
```

NORMSDIST(z)
Returns the standard normal cumulative distribution (has a mean of zero and a standard deviation of one).

Help on this function | OK | Cancel |

4. Complete the NORMSDIST dialog box as shown below. Click **OK**.

Function Arguments ? X

NORMSDIST

Z | -.99 | = -0.99

= 0.16108706

Returns the standard normal cumulative distribution (has a mean of zero and a standard deviation of one).

Z is the value for which you want the distribution.

Formula result = 0.16108706

Help on this function | OK | Cancel |

The output is displayed in cell A1 of the worksheet. The area under the standard normal curve to the left of $z = -0.99$ is 0.1611.

◀

▶ Example 5 (pg. 242)   Finding Area Under the Standard Normal Curve

1. Open a new Excel worksheet. The area to the right of $z = 1.06$ is equal to 1 minus the area to the left of $z = 1.06$. In cell A1, enter **=1-** as shown below.

	A
1	=1-

2. At the top of the screen, click **Formulas** and select **Insert Function**.

3. Select the **Statistical** category. Select the **NORMSDIST** function. Click **OK**.

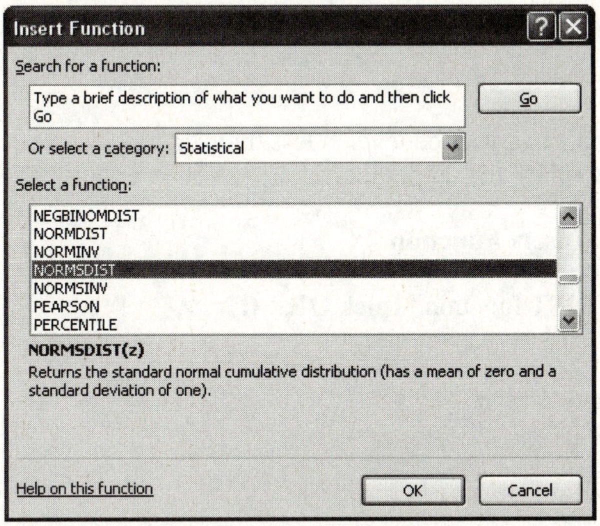

4. Complete the NORMSDIST dialog box as shown below. Click **OK**.

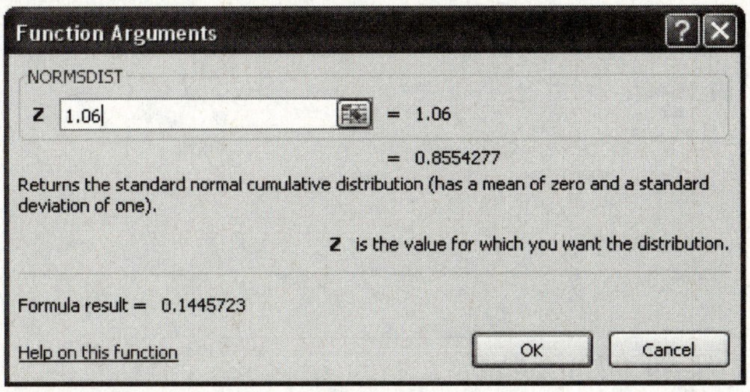

The NORMSDIST function returns a value of 0.8554. The area to the right of $z = 1.06$ is equal to 1- 0.8554. The answer, 0.1446, is displayed in cell A1 of the worksheet. Notice the equation in the formula bar to the right of **f$_x$**, =1-NORMSDIST(1.06).

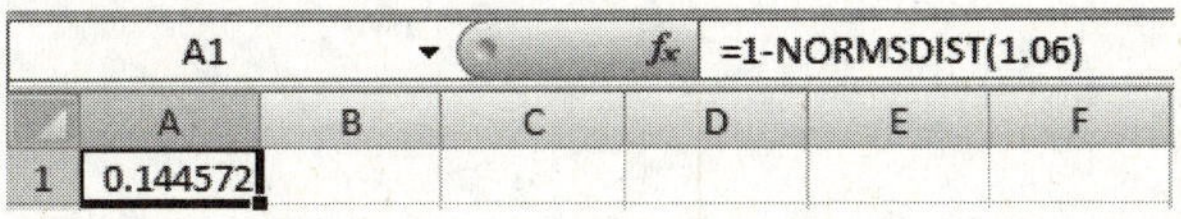

Section 5.2 Normal Distributions: Finding Probabilities

> ▶ Example 3 (pg. 251)   Using Technology to Find Normal Probabilities

1. You are asked to find the probability that the person's triglyceride level is less than 80. Open a new Excel worksheet and click in cell **A1** to place the output there.

2. At the top of the screen, click **Formulas** and select **Insert Function**.

3. Select the **Statistical** category. Select the **NORMDIST** function. Click **OK**.

Insert Function [?] [X]

Search for a function:

Type a brief description of what you want to do and then click Go   Go

Or select a category: Statistical

Select a function:

MINA
MODE
NEGBINOMDIST
NORMDIST
NORMINV
NORMSDIST
NORMSINV

NORMDIST(x,mean,standard_dev,cumulative)
Returns the normal cumulative distribution for the specified mean and standard deviation.

Help on this function   OK   Cancel

Be sure to select NORMDIST and not NORMSDIST for this problem.

4. Complete the NORMDIST dialog box as shown below. Click **OK**.

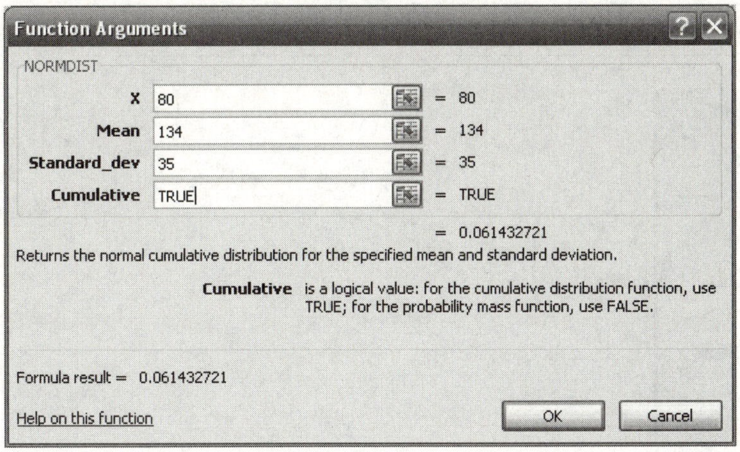

Function Arguments		? X

NORMDIST

X	80		=	80
Mean	134		=	134
Standard_dev	35		=	35
Cumulative	TRUE		=	TRUE

= 0.061432721

Returns the normal cumulative distribution for the specified mean and standard deviation.

Cumulative is a logical value: for the cumulative distribution function, use TRUE; for the probability mass function, use FALSE.

Formula result = 0.061432721

Help on this function   OK   Cancel

The output is displayed in cell A1 of the worksheet. The probability is equal to 0.0614.

◄

5.3 Normal Distributions: Finding Values

► Example 2 (pg. 258)   Finding a *z*-Score Given a Percentile

1. We will start with finding the z-score that corresponds to P_5. Open a new Excel worksheet and click in cell **A1** to place the output there.

2. At the top of the screen, click **Formulas** and select **Insert Function**.

3. Select the **Statistical** category. Select the **NORMSINV** function. Click **OK**.

Be sure to select NORMSINV and not NORMINV.

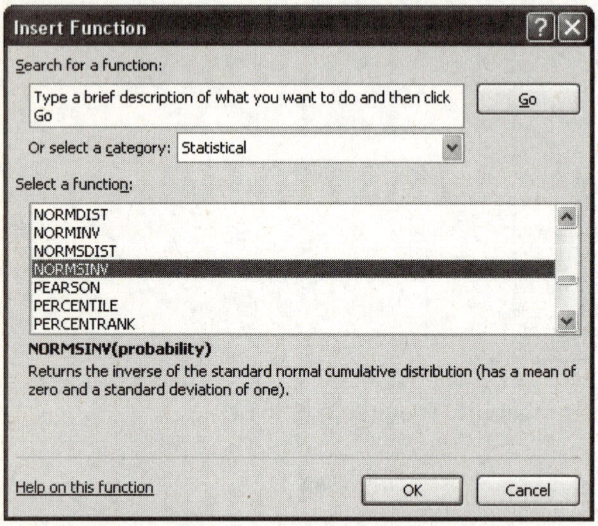

4. Complete the NORMSINV dialog box as shown below. Click **OK**.

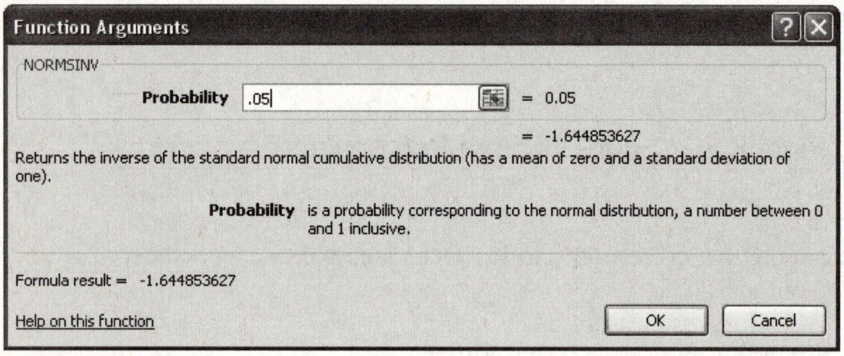

5. The function returns a *z*-score of -1.6449. To find the *z*-scores for P50 and P90, repeat these steps using probabilities of .50 and .90, respectively.

► Example 4 (pg. 260)   Finding a Specific Data Value
_____

1. You will find the lowest score an applicant can earn and still be eligible for employment given that the agency will only hire applicants with scores in the top 10%. Open a new Excel worksheet and click in cell **A1** to place the output there.

2. At the top of the screen, click **Formulas** and select **Insert Function**.

3. Select the **Statistical** category. Select the **NORMINV** function. Click **OK**.

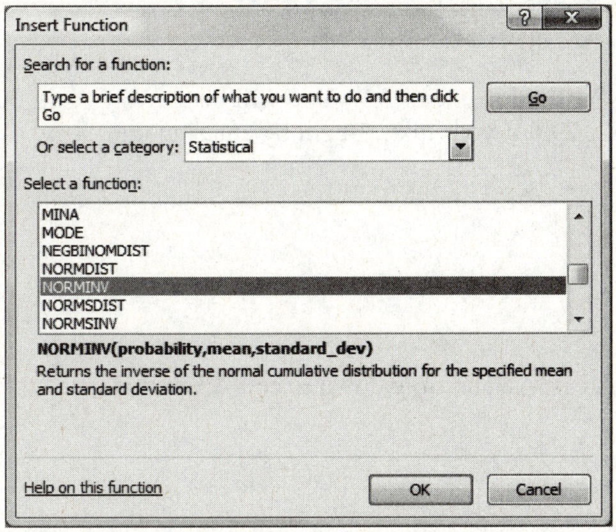

Insert Function	? X

Search for a function:

Type a brief description of what you want to do and then click Go	Go

Or select a category: Statistical ▼

Select a function:

MINA
MODE
NEGBINOMDIST
NORMDIST
NORMINV
NORMSDIST
NORMSINV

NORMINV(probability,mean,standard_dev)
Returns the inverse of the normal cumulative distribution for the specified mean and standard deviation.

Help on this function   OK   Cancel

4. Complete the NORMINV dialog box as shown below. Click **OK**.

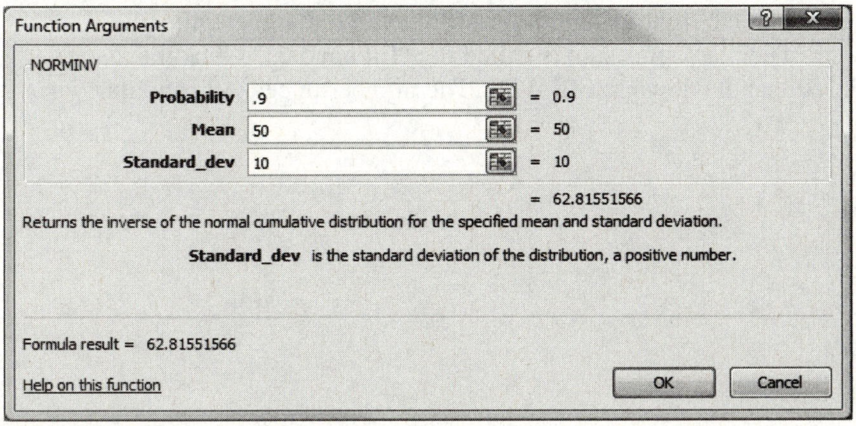

Function Arguments   ? X

NORMINV

Probability	.9	🖩	=	0.9
Mean	50	🖩	=	50
Standard_dev	10	🖩	=	10

= 62.81551566

Returns the inverse of the normal cumulative distribution for the specified mean and standard deviation.

Standard_dev is the standard deviation of the distribution, a positive number.

Formula result = 62.81551566

Help on this function   OK   Cancel

The NORMINV function returns a value of 62.8155.

◄

Technology

► Exercise 1 (pg. 299)	Finding the Mean Age in the United States

1. Open a new, blank Excel worksheet. Enter **Class Midpoint** in cell A1. You will be entering the numbers displayed in the table on page 299 of your text. Begin by entering the first two midpoints, **2** and **7**, as shown below.

	A
1	Class Midpoint
2	2
3	7

2. You will now fill column A with a series. Click in cell A2 and drag down to cell A3 so that both cells are highlighted.

	A
1	Class Midpoint
2	2
3	7

3. Move the mouse pointer in cell A3 to the right lower corner of the cell so that the white plus sign turns into a black plus sign. The black plus sign is called the "fill handle." Click the left mouse button and drag the fill handle down to cell A21. The number in cell A21 should be 97.

	A
1	Class Midpoint
2	2
3	7
4	12
5	17
6	22
7	27
8	32
9	37
10	42
11	47
12	52
13	57
14	62
15	67
16	72
17	77
18	82
19	87
20	92
21	97

4. Enter **Relative Frequency** in cell **B1**. Then enter the proportion equivalents of the percentages in column B as shown below.

	A	B	C
1	Class Midpoint	Relative Frequency	
2	2	0.069	
3	7	0.066	
4	12	0.066	
5	17	0.071	
6	22	0.069	
7	27	0.07	
8	32	0.064	
9	37	0.069	
10	42	0.071	
11	47	0.075	
12	52	0.071	
13	57	0.061	
14	62	0.05	
15	67	0.037	
16	72	0.029	
17	77	0.024	
18	82	0.019	
19	87	0.012	
20	92	0.005	
21	97	0.002	

5. Click in cell **C2** and enter a formula to multiply the midpoint by the relative frequency. The formula is **=A2*B2**. Press [**Enter**].

	A	B	C
1	Class Midpoint	Relative Frequency	
2	2	0.069	=A2*B2

6. Copy the contents of cell C2 to cells C3 through C21.

	A	B	C
1	Class Midpoint	Relative Frequency	
2	2	0.069	0.138
3	7	0.066	0.462
4	12	0.066	0.792
5	17	0.071	1.207
6	22	0.069	1.518
7	27	0.07	1.89
8	32	0.064	2.048
9	37	0.069	2.553
10	42	0.071	2.982
11	47	0.075	3.525
12	52	0.071	3.692
13	57	0.061	3.477
14	62	0.05	3.1
15	67	0.037	2.479
16	72	0.029	2.088
17	77	0.024	1.848
18	82	0.019	1.558
19	87	0.012	1.044
20	92	0.005	0.46
21	97	0.002	0.194

7. The weighted mean is equal to the sum of the products in column C. (Refer to Section 2.3 in your text.) Click in cell **C22** of the worksheet to place the sum there. Click **Formulas** near the top of the screen and select **AutoSum**. The range of numbers to be included in the sum is displayed in cell C22. You should see **=SUM(C2:C21)**. Make any necessary corrections. Then press [**Enter**].

21	97	0.002	0.194
22			=SUM(C2:C21)

8. The mean, displayed in cell C22, is 37.055 years. Save this worksheet so that you can use it again for Exercise 5, page 299.

◄

► Exercise 2 (pg. 299)   Finding the Mean of the Set of Sample Means

1.   Open worksheet "Tech5" in the Chapter 5 folder.

2.   Click in cell **A38** at the bottom of the column of numbers to place the mean in that cell.

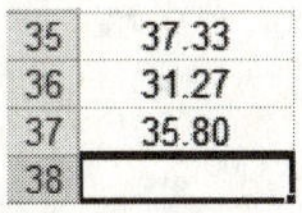

35	37.33
36	31.27
37	35.80
38	

3.   At the top of the screen, click **Formulas** and select **Insert Function**.

4.   Select the **Statistical** category. Select the **AVERAGE** function. Click **OK**.

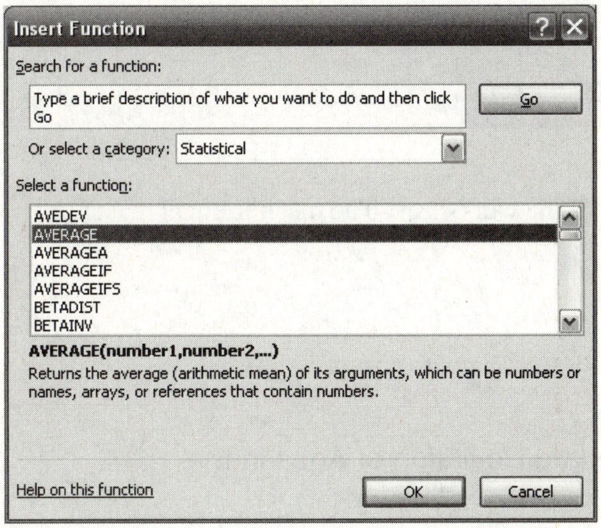

Insert Function

Search for a function:

Type a brief description of what you want to do and then click Go   Go

Or select a category: Statistical

Select a function:

AVEDEV
AVERAGE
AVERAGEA
AVERAGEIF
AVERAGEIFS
BETADIST
BETAINV

AVERAGE(number1,number2,...)
Returns the average (arithmetic mean) of its arguments, which can be numbers or names, arrays, or references that contain numbers.

Help on this function   OK   Cancel

5. The range that will be included in the average is shown in the Number 1 window. Check to be sure that it is accurate. It should read **A2:A37**. Make any necessary corrections. Then click **OK**.

Function Arguments		? ✕
AVERAGE		
Number1	A2:A37 ▦	= {28.14;31.56;36.86;32.37;36.12;39...
Number2	▦	= number
		= 36.20944444

Returns the average (arithmetic mean) of its arguments, which can be numbers or names, arrays, or references that contain numbers.

      Number1: number1,number2,... are 1 to 255 numeric arguments for which you want the average.

Formula result = 36.20944444

Help on this function   OK   Cancel

The mean of the sample means is displayed in cell A38 of the worksheet. The mean is equal to 36.21.

◀

▶ Exercise 5 (pg. 299)   Finding the Standard Deviation of Ages in the United States

1. Open the worksheet that you prepared for Exercise 1 on page 299. The first few lines are shown below.

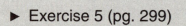

	A	B	C
1	Class Midpoint	Relative Frequency	
2	2	0.069	0.138
3	7	0.066	0.462
4	12	0.066	0.792
5	17	0.071	1.207

2. To find the standard deviation of the population of ages, you will first calculate squared deviation scores. Click in cell **D1** and enter the label, **Sqd Dev Score**.

	A	B	C	D	E
1	Class Midpoint	Relative Frequency	Sqd Dev Score		
2	2	0.069	0.138		
3	7	0.066	0.462		

3. Click in cell **D2** and enter the formula **=(A2-37.055)^2** and press [**Enter**].

	A	B	C	D	E
1	Class Midpoint	Relative Frequency	Sqd Dev Score		
2	2	0.069	0.138	=(A2-37.055)^2	
3	7	0.066	0.462		

4. Copy the formula in cell D2 to cells D3 through D21.

	A	B	C	D	E
1	Class Midpoint	Relative Frequency	Sqd Dev Score		
2	2	0.069	0.138	1228.853	
3	7	0.066	0.462	903.303	
4	12	0.066	0.792	627.753	
5	17	0.071	1.207	402.203	
6	22	0.069	1.518	226.653	
7	27	0.07	1.89	101.103	
8	32	0.064	2.048	25.55303	
9	37	0.069	2.553	0.003025	
10	42	0.071	2.982	24.45303	
11	47	0.075	3.525	98.90303	
12	52	0.071	3.692	223.353	
13	57	0.061	3.477	397.803	
14	62	0.05	3.1	622.253	
15	67	0.037	2.479	896.703	
16	72	0.029	2.088	1221.153	
17	77	0.024	1.848	1595.603	
18	82	0.019	1.558	2020.053	
19	87	0.012	1.044	2494.503	
20	92	0.005	0.46	3018.953	
21	97	0.002	0.194	3593.403	

5. Each of the squared deviations will be weighted by its relative frequency. Click in cell **E2**. Enter the formula **=D2*B2** and press [**Enter**].

	A	B	C	D	E
1	Class Midpoint	Relative Frequency	Sqd Dev Score		
2	2	0.069	0.138	1228.853	=D2*B2
3	7	0.066	0.462	903.303	

6. Copy the formula in cell E2 to cells E3 through E21.

	A	B	C	D	E
1	Class Midpoint	Relative Frequency		Sqd Dev Score	
2	2	0.069	0.138	1228.853	84.79086
3	7	0.066	0.462	903.303	59.618
4	12	0.066	0.792	627.753	41.4317
5	17	0.071	1.207	402.203	28.55641
6	22	0.069	1.518	226.653	15.63906
7	27	0.07	1.89	101.103	7.077212
8	32	0.064	2.048	25.55303	1.635394
9	37	0.069	2.553	0.003025	0.000209
10	42	0.071	2.982	24.45303	1.736165
11	47	0.075	3.525	98.90303	7.417727
12	52	0.071	3.692	223.353	15.85806
13	57	0.061	3.477	397.803	24.26598
14	62	0.05	3.1	622.253	31.11265
15	67	0.037	2.479	896.703	33.17801
16	72	0.029	2.088	1221.153	35.41344
17	77	0.024	1.848	1595.603	38.29447
18	82	0.019	1.558	2020.053	38.38101
19	87	0.012	1.044	2494.503	29.93404
20	92	0.005	0.46	3018.953	15.09477
21	97	0.002	0.194	3593.403	7.186806

7. You are working with a relative frequency distribution of midpoints. For this type of distribution, the variance is equal to the sum of the squared deviation scores. Click in cell **E22** to place the sum of the squared deviation scores there. Click **Formulas** near the top of the screen and select **AutoSum.** The range of numbers to be included in the sum is displayed in cell E22. You should see **=SUM(E2:E21)**. Make any necessary corrections. Then press [**Enter**].

21		97	0.002	0.194	3593.403	7.186806
22				37.055		=SUM(E2:E21)

8. Click in cell **E23** to place the standard deviation there. The standard deviation is the square root of the variance. In cell E23, enter the formula **=sqrt(E22)** and press [**Enter**].

21		97	0.002	0.194	3593.403	7.186806
22				37.055		516.622
23						=sqrt(E22)

9. The standard deviation of ages in the United States is approximately 22.73. For future reference purposes, you may want to add labels for the mean, variance, and standard deviation as shown below.

| 22 | | Mean = | 37.055 | Var = | 516.622 |
| 23 | | | | St Dev = | 22.72932 |

◀

▶ Exercise 6 (pg. 299)	Finding the Standard Deviation of the Set of Sample Means

1. Open worksheet "Tech5" in the Chapter 5 folder. This is the same data set that you used for Exercise 2, page 299. The first few lines are shown below. If you placed the mean (36.21) in cell A38 for Exercise 2 on page 299, you should delete it now.

	A
1	Mean ages
2	28.14
3	31.56
4	36.86
5	32.37

2. At the top of the screen, select **Data** and **Data Analysis**.

If Data Analysis does not appear as a choice in the Data ribbon, you will need to load the Microsoft Excel Analysis ToolPak add-in. Follow the procedure in Section GS 8.1 before continuing.

3. Select **Descriptive Statistics** and click **OK**.

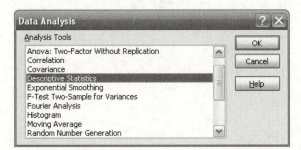

4. Complete the Descriptive Statistics dialog box as shown at the top of the next page. Be sure to select **Labels in First Row** near the top of the dialog box and **Summary statistics** near the bottom. Click **OK**.

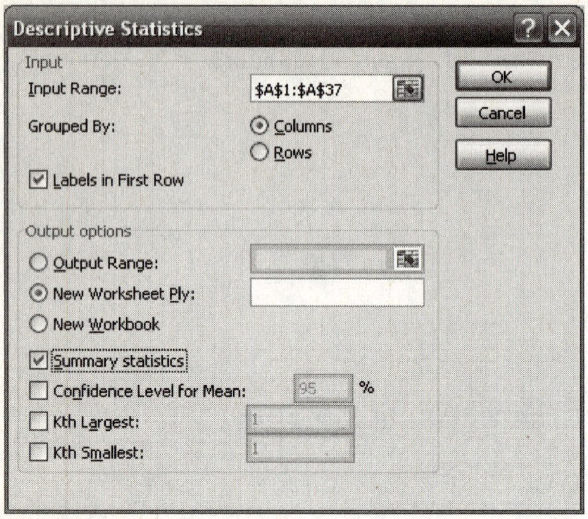

Descriptive Statistics

Input

Input Range: $A$1:$A$37

Grouped By:
- ◉ Columns
- ○ Rows

☑ Labels in First Row

Output options

- ○ Output Range:
- ◉ New Worksheet Ply:
- ○ New Workbook

☑ Summary statistics
☐ Confidence Level for Mean: 95 %
☐ Kth Largest: 1
☐ Kth Smallest: 1

OK
Cancel
Help

You will want to make column A wider so that you can read all the labels in the output table. The standard deviation is equal to 3.5518.

	A	B
1	*Mean ages*	
2		
3	Mean	36.20944
4	Standard Error	0.591967
5	Median	36.155
6	Mode	#N/A
7	Standard Deviation	3.551804
8	Sample Variance	12.61531
9	Kurtosis	0.343186
10	Skewness	0.207283
11	Range	16.58
12	Minimum	28.14
13	Maximum	44.72
14	Sum	1303.54
15	Count	36

Confidence Intervals

Section 6.1 Confidence Intervals for the Mean (Large Samples)

► Example 4 (pg. 308)	Constructing a Confidence Interval Using Technology

If the DDXL add-in has not been loaded, you will need to load it before continuing. Follow the instructions in Section GS 8.2.

1. Open the "Friends" worksheet in the Chapter 6 folder.

2. Click and drag over the data range, **A1:A41**, so that it is highlighted.

3. At the top of the screen, select **Add-Ins** and **DDXL**. Select **Confidence Intervals.**

4. Click the down arrow under Function type and select **1 Var z Interval**. Note that there should be a checkmark in the box next to First row is variable names.

5. Select **Friends** in the Names and Columns window. Then click the arrow under Quantitative Variable on the left to select Friends for the confidence interval. Click **OK**.

6. Click **99%** to select a 99% confidence interval. Type **53** in the window provided for the population standard deviation. Click **Compute Interval**.

The output is displayed below. The lower limit of the 99% confidence interval is 109.214 and the upper limit is 152.386.

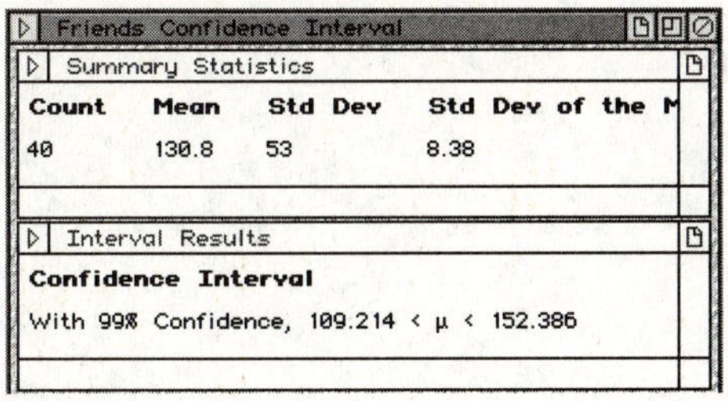

```
┌──────────────────────────────────────────┐
│ ▷│ Friends Confidence Interval   🗎🗔⊘    │
│ ┌────────────────────────────────────┐🗎 │
│ │▷│ Summary Statistics               │  │
│ │ Count   Mean   Std Dev   Std Dev of the M│
│ │ 40      130.8   53        8.38          │
│ │                                        │
│ │                                        │
│ └────────────────────────────────────┘  │
│ ┌────────────────────────────────────┐🗎 │
│ │▷│ Interval Results                 │  │
│ │ Confidence Interval                   │
│ │ With 99% Confidence, 109.214 < μ < 152.386│
│ └────────────────────────────────────┘  │
│                                          │
└──────────────────────────────────────────┘
```

◀

▶ Example 5 (pg. 309)	Constructing a Confidence Interval, σ Known

1. Open a new Excel worksheet. At the top of the screen, select **Formulas** and **Insert Function**.

2. Select the **Statistical** category and the **CONFIDENCE** function. Click **OK**.

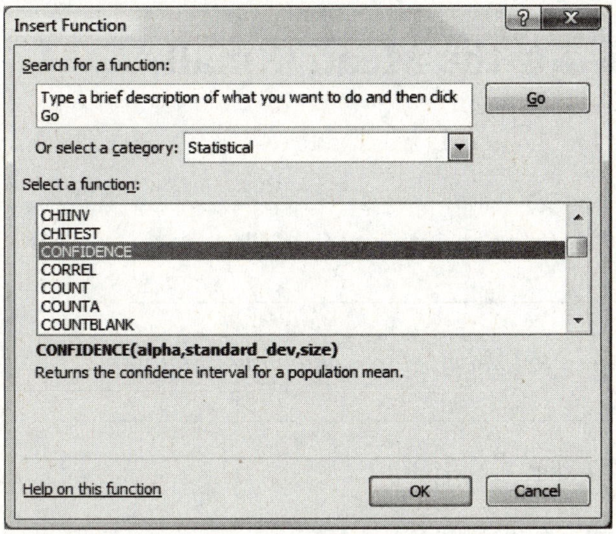

```
┌────────────────────────────────────────────────┐
│ Insert Function                        ?  X     │
│                                                 │
│ Search for a function:                          │
│ ┌──────────────────────────────────┐ ┌───────┐ │
│ │ Type a brief description of what  │ │  Go   │ │
│ │ you want to do and then click Go  │ └───────┘ │
│ └──────────────────────────────────┘           │
│ Or select a category: │ Statistical    ▼│       │
│                                                 │
│ Select a function:                              │
│ ┌────────────────────────────────────────┐▲    │
│ │ CHIINV                                  │     │
│ │ CHITEST                                 │     │
│ │ CONFIDENCE                              │     │
│ │ CORREL                                  │     │
│ │ COUNT                                   │     │
│ │ COUNTA                                  │     │
│ │ COUNTBLANK                              │▼    │
│ └────────────────────────────────────────┘     │
│ CONFIDENCE(alpha,standard_dev,size)             │
│ Returns the confidence interval for a population mean.│
│                                                 │
│                                                 │
│ Help on this function      ┌────┐  ┌────────┐   │
│                            │ OK │  │ Cancel │   │
│                            └────┘  └────────┘   │
└────────────────────────────────────────────────┘
```

3. Complete the dialog box as shown below. Alpha is equal to .10, the standard deviation is equal to 1.5, and the sample size is 20. Click **OK**.

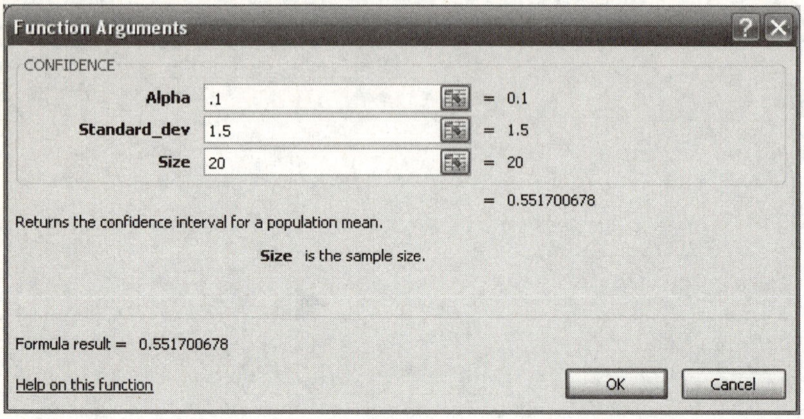

Function Arguments	? ✕

CONFIDENCE

Alpha	.1	📷	= 0.1
Standard_dev	1.5	📷	= 1.5
Size	20	📷	= 20

= 0.551700678

Returns the confidence interval for a population mean.

      Size is the sample size.

Formula result = 0.551700678

Help on this function   [OK] [Cancel]

4. The function returns a result of 0.552. You need to do two calculations to obtain the confidence interval. The left endpoint of the confidence interval is equal to (22.9 – 0.552), and the right endpoint is equal to (22.9 + 0.552). With 90% confidence, you can say that the mean age of all the students is between 22.348 and 23.452 years.

◀

Section 6.2 Confidence Intervals for the Mean (Small Samples)

▶ Exercise 29 (pg. 325)	Constructing a Confidence Interval for Miles per Gallon

If the DDXL add-in has not been loaded, you will need to load it before continuing. Follow the instructions in Section GS 8.2.

1. Open the "Ex6_2-29" worksheet in the Chapter 6 folder.

2. Click and drag over the data range, **A1:A26**, so that it is highlighted.

3. At the top of the screen, select **Add-Ins** and **DDXL**. Select **Confidence Intervals.**

4. Click the down arrow under Function type and select **1 Var t Interval**. Note that there should be a checkmark in the box next to First row is variable names.

5. Select **Miles per gallon** in the Names and Columns window. Then click the arrow under Quantitative Variable on the left to select Miles per gallon for the confidence interval. Click **OK**.

6. Click **95%** to select a 95% confidence interval. Click **Compute Interval**.

The output is shown at the top of the next page. We can say with 95% confidence that the population mean is between 20.49 and 23.35 miles per gallon.

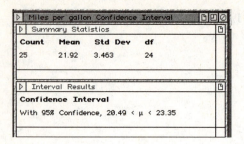

Section 6.3 Confidence Intervals for Population Proportions

▶ **Example 2 (pg. 329)**   Constructing a Confidence Interval for *p*

The problem asks you to construct a 95% confidence interval for the proportion of adults in the United States who say it is acceptable to check personal e-mail while at work. In a survey of 1,000 U.S. adults, 662 said it was acceptable.

If the DDXL add-in has not been loaded, you will need to load it before continuing. Follow the instructions in Section GS 8.2.

1. Open a new Excel worksheet. In cell A1, enter **662**, the number of successes. In cell B1, enter **1000**, the sample size. Click and drag over cells A1 and B1 so that both are highlighted.

	A	B
1	662	1000

2. At the top of the screen, select **Add-Ins** and **DDXL**. Select **Confidence Intervals.**

Summaries
Tables
Charts and Plots
Regression
ANOVA
Confidence Intervals
Exponential Smoothing

3. Click the down arrow under Function type and select **Summ 1 Var Prop Interval**.

Confidence Intervals Dialog

Function type:

Summ 1 Var Prop Interva ▼

This command computes a confidence interval for a proportion for summarized data. You need one column that holds the number of success and one column that holds the number of trials. There

1 Var Prop Interval
1 Var t Interval
1 Var z Interval
2 Var Prop Interval
2 Var t Interval
Chi-square Conf Ints for SD
F Conf Ints of SD
Paired t Interval
Summ 1 Var Prop Interval
Summ 2 Var Prop Interval

Names and Columns

$A$1
$B$1

Num Trials

☐ First row is variable names

Info   Help

Cancel   OK

4. Select **$A$1** in the Names and Columns window. Then click the arrow under Num Successes on the left to select the contents of cell A1 for the number of successes.

5. Select **$B$1** in the Names and Columns window. Then click the arrow under Num Trials on the left to select the contents of cell B1 for the number of trials. The completed dialog box is shown at the top of the next page. Click **OK**.

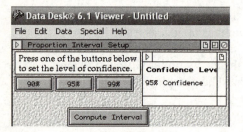

6. Click **95%** to select a 95% confidence interval. Click **Compute Interval**.

Your output should look like the output displayed below. The left endpoint of the 95% confidence interval is 0.633 and the right endpoint is 0.691.

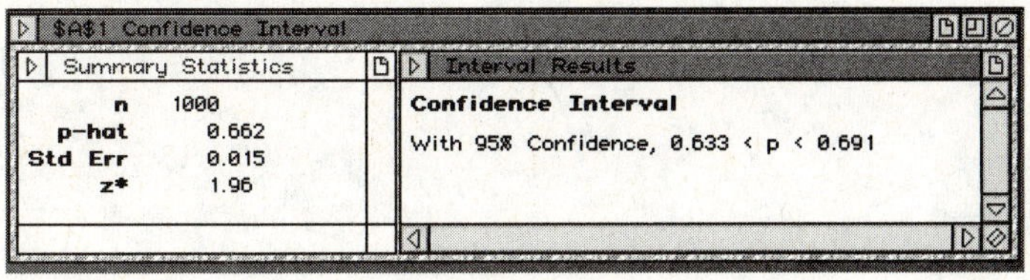

```
┌─▷─ $A$1 Confidence Interval ──────────────────────────[▯▯▯]─┐
│ ┌─▷─ Summary Statistics ──[▯]┐┌─▷─ Interval Results ──────[▯]─┐
│ │                          ││  Confidence Interval        ▲  │
│ │      n    1000           ││                             ░  │
│ │  p-hat    0.662          ││  With 95% Confidence, 0.633 < p < 0.691 │
│ │ Std Err   0.015          ││                             ▽  │
│ │     z*    1.96           ││                             ░  │
│ │                          ││                             ░  │
│ └──────────────────────────┘└─◁─────────────────────▷─◈─┘  │
└──────────────────────────────────────────────────────────────┘
```

◀

6.4 Confidence Intervals for Variation and Standard Deviation

▶ Exercise 11 (pg. 341)	Constructing a 99% Confidence Interval

If the DDXL add-in has not been loaded, you will need to load it before continuing. Follow the instructions in Section GS 8.2.

1. Open a new Excel worksheet. Enter the reserve capacities of the 18 car batteries as shown below.

	A
1	Hours
2	1.7
3	1.6
4	1.94
5	1.58
6	1.74
7	1.6
8	1.86
9	1.72
10	1.38
11	1.46
12	1.64
13	1.49
14	1.55
15	1.7
16	1.75
17	0.88
18	1.77
19	2.07

2. Click and drag over the data range **A1:A19** so that it is highlighted. At the top of the screen, select **Add-Ins** and **DDXL**. Select **Confidence Intervals**

3. Select **Chi-square Conf Ints for SD**. Note that there should be a checkmark in the box to the left of First row is variable names.

4. Select **Hours** in the Names and Columns window. Then click the arrow under Quantitative Variable on the left to select Hours for the confidence interval. Click **OK**.

5. Click **99%** to obtain the 99% confidence interval for the standard deviation.

```
┌──────────────────────────────────────────────────────────────┐
│ ▷ Compute Chi-square Confidence Ints for the Std Dev: scores ☐▣⊘│
│           ┌─────────────────────────────────┐                  │
│           │  Press one of the buttons below │                  │
│           │  to set the level of confidence.│                  │
│           └─────────────────────────────────┘                  │
│              ┌─────┐   ┌─────┐   ┌─────┐                        │
│              │ 90% │   │ 95% │   │ 99% │                        │
│              └─────┘   └─────┘   └─────┘                        │
└──────────────────────────────────────────────────────────────┘
```

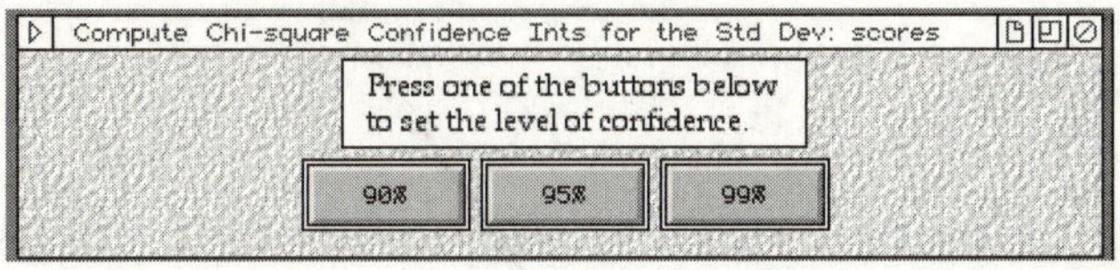

The output is shown below. The standard deviation is 0.253. The lower limit of the 99% confidence interval is 0.175 and the upper limit is 0.437.

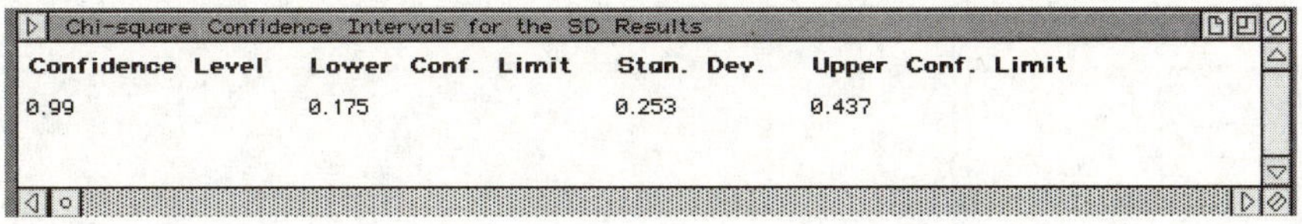

```
┌────────────────────────────────────────────────────────────────┐
│ ▷ Chi-square Confidence Intervals for the SD Results       ☐▣⊘  │
│ Confidence Level    Lower Conf. Limit    Stan. Dev.   Upper Conf. Limit│
│ 0.99                0.175                0.253        0.437      │
└────────────────────────────────────────────────────────────────┘
```

Technology

▶ Exercise 1 (pg. 351) Using Technology to Find a 95% Confidence Interval

If the DDXL add-in has not been loaded, you will need to load it before continuing. Follow the instructions in Section GS 8.2.

You are told that 30% of the adults named Barack Obama and that the sample size was 1,025. To use DDXL to compute a confidence interval, you will need the number of successes. The number of successes is 30% of 1,025 or 308.

1. Open a new Excel worksheet. Enter **308** in cell A1 and **1025** in cell B1. Click and drag over the range **A1:B1** so that the cells are highlighted.

	A	B
1	308	1025

2. At the top of the screen, select **Add-Ins** and **DDXL**. Select **Confidence Intervals**.

3. Select **Summ 1 Var Prop Interval**. Note that there should not be a checkmark in the box to the left of First row is variable names.

4. Select **$A$1** in the Names and Columns window. Then click the arrow under Num Successes on the left to select the contents of cell A1 for the number of successes.

5. Select **$B$1** in the Names and Columns window. Then click the arrow under Num Trials on the left to select the contents of cell B1 for the number of trials. The completed dialog box is shown at the top of the next page. Click **OK**.

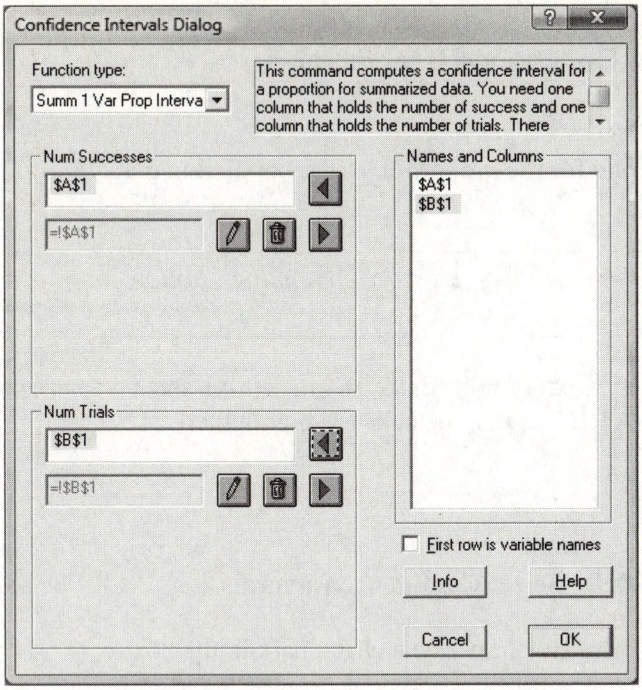

6.   Click **95%** to select a 95% confidence interval. Click **Compute Interval**.

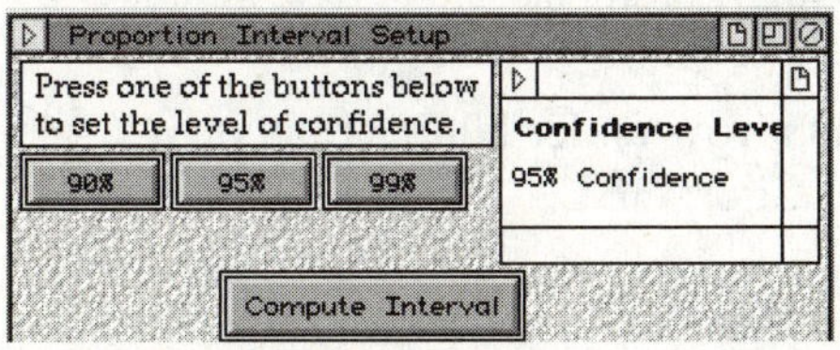

Your output should look the same as the output displayed below. The lower limit of the 95% confidence interval is 0.272 and the upper limit is 0.329.

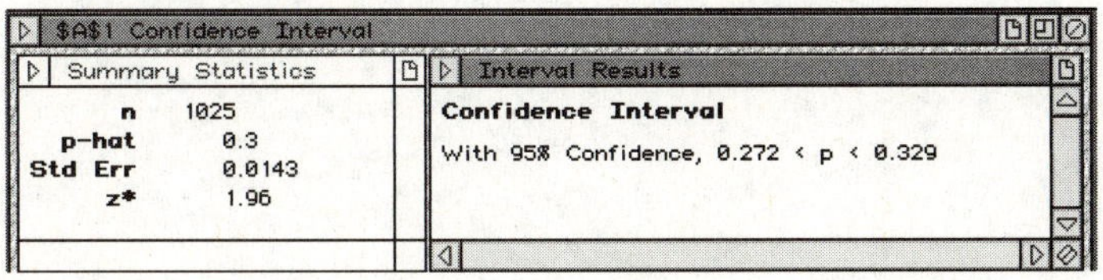

> ► Exercise 2 (pg. 351)   Finding a 95% Confidence Interval for *p*

The problem asks you to find a 95% confidence interval for the proportion of the population that would have chosen Hillary Clinton.

If the DDXL add-in has not been loaded, you will need to load it before continuing. Follow the instructions in Section GS 8.2.

---

1. Open a new Excel worksheet. Sixteen percent of 1,025 is 164. Enter **164** in cell A1 and **1025** in cell B1. Click and drag over the range **A1:B1** so that the cells are highlighted.

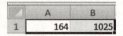

	A	B
1	164	1025

2. At the top of the screen, select **Add-Ins** and **DDXL**. Select **Confidence Intervals**.

3. Select **Summ 1 Var Prop Interval**. Note that there should not be a checkmark in the box to the left of First row is variable names.

4. Select **$A$1** in the Names and Columns window. Then click the arrow under Num Successes on the left to select the contents of cell A1 for the number of successes.

5. Select **$B$1** in the Names and Columns window. Then click the arrow under Num Trials on the left to select the contents of cell B1 for the number of trials. Click **OK**.

6. Click **95%** to select a 95% confidence interval. Click **Compute Interval**.

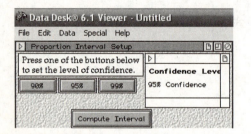

Your output should look the same as the output displayed below. The lower limit of the 95% confidence interval is 0.138 and the upper limit is 0.182.

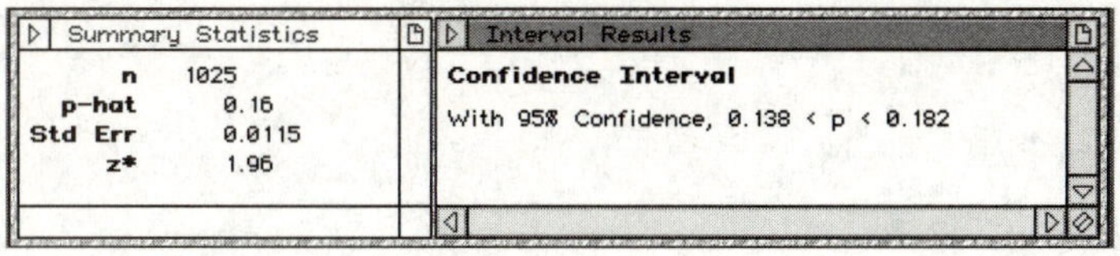

▷	Summary Statistics		▷	Interval Results	
	n	1025		**Confidence Interval**	
	p-hat	0.16		With 95% Confidence, 0.138 < p < 0.182	
	Std Err	0.0115			
	z*	1.96			

◀

▶ Exercise 5 (pg. 351)	Simulating a Most Admired Poll

1. Open a new, blank Excel worksheet. Enter the labels shown below for displaying the output of five simulations.

	A
1	Time 1
2	Time 2
3	Time 3
4	Time 4
5	Time 5

2. At the top of the screen, select **Data** and **Data Analysis**. Select **Random Number Generation**. Click **OK**.

If Data Analysis does not appear as a choice in the Data ribbon, you will need to load the Microsoft Excel Analysis ToolPak add-in. Follow the procedure in Section GS 8.1 before continuing.

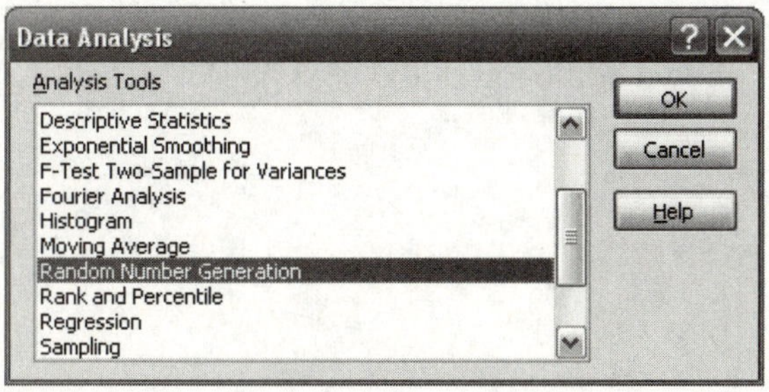

Data Analysis   ? ✕

Analysis Tools

Descriptive Statistics
Exponential Smoothing
F-Test Two-Sample for Variances
Fourier Analysis
Histogram
Moving Average
Random Number Generation
Rank and Percentile
Regression
Sampling

OK
Cancel
Help

3. Complete the Random Number Generation dialog box as shown below. The output will be placed in cell B1 of the worksheet. Click **OK**.

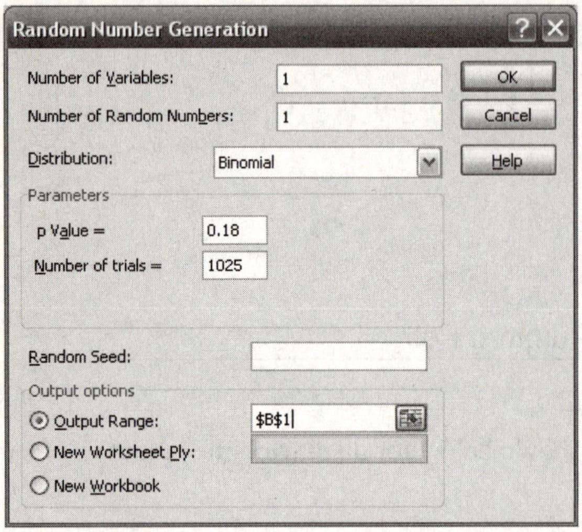

Random Number Generation	? X
Number of Variables: `1`	OK
Number of Random Numbers: `1`	Cancel
Distribution: `Binomial` ▼	Help
Parameters	
p Value = `0.18`	
Number of trials = `1025`	
Random Seed:	
Output options	
⦿ Output Range: `$B$1`	
○ New Worksheet Ply:	
○ New Workbook	

For Time 1, the number of persons (out of 1,025) selecting Sarah Palin is 190. Because this number was generated randomly, it is unlikely that your number is the same.

	A	B
1	Time 1	190
2	Time 2	
3	Time 3	
4	Time 4	
5	Time 5	

4. Begin Time 2 the same way that you did Time 1. At the top of the screen, select **Data** and **Data Analysis**. Select **Random Number Generation**. Click **OK**.

5. Complete the Random Number Generation dialog box in the same way as you did for Time 1 except change the location of the output to cell B2 as shown at the top of the next page. Click **OK**.

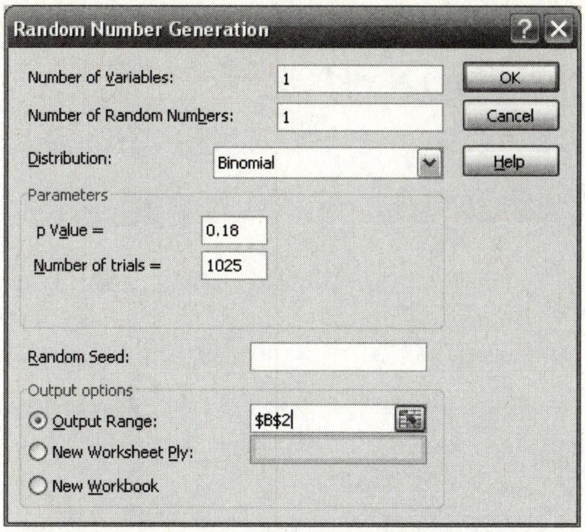

Random Number Generation   ? X

Number of Variables:	1
Number of Random Numbers:	1
Distribution:	Binomial

OK

Cancel

Help

Parameters

p Value = 0.18

Number of trials = 1025

Random Seed:

Output options

⊙ Output Range: $B$2

○ New Worksheet Ply:

○ New Workbook

6. Repeat this procedure until you have carried out the simulation five times. The output I generated is displayed below.

	A	B
1	Time 1	190
2	Time 2	177
3	Time 3	213
4	Time 4	191
5	Time 5	177

◄

Hypothesis Testing with One Sample

Section 7.2 Hypothesis Testing for the Mean (Large Samples)

► Example 2 (pg. 372) | Finding a *P*-Value for a Left-Tailed Test

1. Open a new Excel worksheet and click in cell **A1** to place the output there.

2. At the top of the screen, select **Formulas** and **Insert Function**.

3. Select the **Statistical** category and the **NORMSDIST** function. Click **OK**.

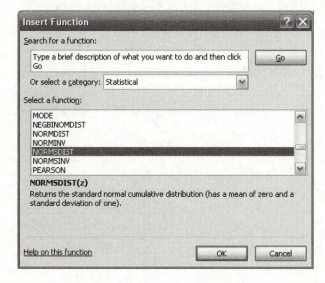

4.   Complete the dialog box as shown below to find the area to the left of $z = -2.23$. Click **OK**.

The function returns a result of 0.0129.

◄

► Example 3 (pg. 372)	Finding a *P*-Value for a Two-Tailed Test

1.   Open a new Excel worksheet and click in cell **A1** to place the output there.

2.   You will be using the NORMSDIST function to find the one-tailed probability. You will then multiply the one-tailed probability by 2 to find the two-tailed probability. So, begin by entering =2* as shown below.

	A
1	=2*

3.   At the top of the screen, select **Formulas** and **Insert Function**.

4.   Select the **Statistical** category and the **NORMSDIST** function. Click **OK**.

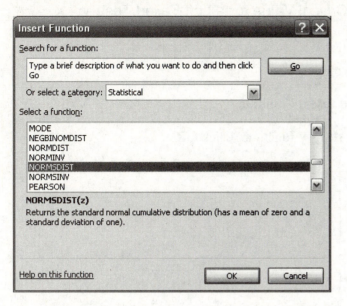

5.   Complete the dialog box as shown below to find the area to the left of $z = -2.14$. The area
     will be multiplied by 2. Click **OK**.

The formula result is 0.0324.

▶ Example 7 (pg. 377)   Finding a Critical Value for a Left-Tailed Test

1. Click in the cell where you want to place the output.

2. At the top of the screen, select **Formulas** and **Insert Function**.

3. Select the **Statistical** category and the **NORMSINV** function. Click **OK**.

Insert Function

Search for a function:

Type a brief description of what you want to do and then click Go   [Go]

Or select a category: Statistical

Select a function:

```
NEGBINOMDIST
NORMDIST
NORMINV
NORMSDIST
NORMSINV
PEARSON
PERCENTILE
```

NORMSINV(probability)
Returns the inverse of the standard normal cumulative distribution (has a mean of zero and a standard deviation of one).

Help on this function   [OK]   [Cancel]

4. You want to find the critical value for an area of 0.01 in the left tail of the standard normal curve. So, type **.01** in the probability window of the dialog box. Click **OK**.

Function Arguments

NORMSINV

Probability .01   = 0.01

                                              = -2.326347874

Returns the inverse of the standard normal cumulative distribution (has a mean of zero and a standard deviation of one).

              Probability is a probability corresponding to the normal distribution, a number between 0 and 1 inclusive.

Formula result = -2.326347874

Help on this function   [OK]   [Cancel]

The function returns a result of -2.33.

◄

| ► Example 8 (pg. 377) | Finding a Critical Value for a Two-Tailed Test |

1. Click in the cell where you want to place the output.

2. You will be using the NORMSINV function to find the critical value for a two-tailed test with $\alpha = 0.05$. α is distributed equally between the left and right tails of the distribution for a two-tailed test. So, in order to find the two-tailed critical value using NORMSINV, you will be entering $\frac{1}{2}$ α or 0.025.

3. At the top of the screen, select **Formulas** and **Insert Function**.

4. Select the **Statistical** category and the **NORMSINV** function. Click **OK**.

5.   You want to find the critical value for an area of 0.025 in each tail of the standard normal curve. To obtain this value, type **.025** in the probability window of the dialog box. Click **OK**.

The function returns -1.96, the critical value for the left tail. The critical value for the right tail is 1.96.

| ▶ Exercise 33 (pg. 383) | Testing That the Mean Time to Quit Smoking Is 15 Years |

If the DDXL add-in has not been loaded, you will need to load it before continuing. Follow the instructions in Section GS 8.2.

1.   Open the "Ex7_2-33" worksheet in the Chapter 7 folder.

2.   Your textbook indicates that the standard deviation of a sample may be used in the z-test formula instead of a known population standard deviation if the sample size is 30 or greater. You will use Excel's STDEV function to obtain the sample standard deviation of the years data. Click the cell where you want to place the standard deviation.

3.   At the top of the screen, click **Formulas** and select **Insert Function**.

4.   Select the **Statistical** category. Select the **STDEV** function. Click **OK**.

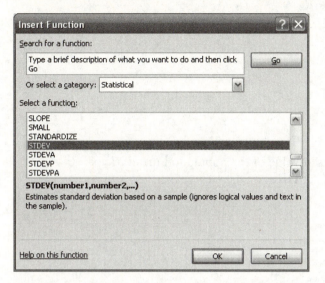

5.   Enter the data range in the Number 1 window as shown below. Click **OK**.

Function Arguments

STDEV

Number1 A1:A33 = {"Years";15.7;13.2;22.6;13;10.7;18....
Number2   = number

= 4.287612267

Estimates standard deviation based on a sample (ignores logical values and text in the sample).

Number1: number1,number2,... are 1 to 255 numbers corresponding to a sample of a population and can be numbers or references that contain numbers.

Formula result = 4.287612267

Help on this function   OK   Cancel

6.   The function returns a result of 4.2876. You will use this value for the population standard deviation when you carry out the statistical test. In the worksheet, click and drag over the data range, **A1:A33**, so that it is highlighted.

7.   At the top of the screen, click **Add-Ins** and select **DDXL**. Select **Hypothesis Tests**.

8.   Click the down arrow under Function type and select **1 Var *z* Test**.

9. Click **Years** in the Names and Columns window. Click the arrow below Quantitative Variable to select Years for the analysis. Click **OK**.

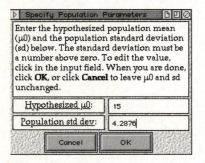

10. Click **Set μ0 and sd**. Type **15** for the hypothesized value of the population mean. Type **4.2876** for the population standard deviation. Click **OK**.

```
▷ | Specify Population Parameters      ⊡⊡∅
  Enter the hypothesized population mean
  (μ0) and the population standard deviation
  (sd) below. The standard deviation must be
  a number above zero. To edit the value,
  click in the input field. When you are done,
  click OK, or click Cancel to leave μ0 and sd
  unchanged.

    Hypothesized μ0:     15

    Population std dev:   4.2876

         Cancel          OK
```

11. Alpha has already been set at 0.05, so you do not need to select the significance level. For the alternative hypothesis, select $\mu \neq \mu_0$. Click **Compute**.

Your output should look like the output displayed below. The *z* statistic is -0.22 and the *p*-value is 0.827.

```
▷ | Years z Test                                          ⊡⊡∅
 ▷ | Summary Statistics    ⊡ |▷ | Test Summary              ⊡
        Count    32                  Ho:            μ = 15
        Mean     14.834              Ha:   2-tailed: μ ≠ 15
    Pop StDev:   4.288        z Statistic:            -0.22
                                  p-value:             0.827

                               ▷ | Test Results             ⊡
                               Conclusion
                               Fail to reject Ho at alpha = 0.05

                          New Test
```

◀

Section 7.3 Hypothesis Testing for the Mean (Small Samples)

▶ Example 1 (pg. 388)   Finding Critical Values for *t*

1. Open a new Excel worksheet and click in cell **A1** to place the output there.

2. At the top of the screen, select **Formulas** and **Insert Function**.

3. Select the **Statistical** category and the **TINV** function. Click **OK**.

4. You want to find the critical value for α equal to 0.05 in the left tail of a t-distribution with 20 degrees of freedom.   Because the TINV function returns a two-tailed critical value, you need to multiply the one-tailed α by 2 to obtain the desired one-tailed critical value. For this example, you multiply 0.05 by 2. The result is 0.10. Type **.10** in the probability window of the dialog box. Type **20** in the Deg_freedom window. Click **OK**.

Function Arguments	? X
TINV	
Probability .10 = 0.1	
Deg_freedom 20 = 20	
= 1.724718218	
Returns the inverse of the Student's t-distribution.	
Deg_freedom is a positive integer indicating the number of degrees of freedom to characterize the distribution.	
Formula result = 1.724718218	
Help on this function	OK Cancel

The result is 1.7247.

◀

▶ Exercise 30 (pg. 396)   Testing the Claim That the Mean Number of Hours Is 11.0

If the DDXL add-in has not been loaded, you will need to load it before continuing. Follow the instructions in Section GS 8.2.

1. Enter the data in an Excel worksheet as shown below.

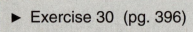

	A
1	Hours
2	11.8
3	8.6
4	12.6
5	7.9
6	6.4
7	10.4
8	13.6
9	9.1

2. Click and drag over the range **A1:A9** to select the data for the analysis.

3. At the top of the screen, click **Add-Ins** and select **DDXL**. Select **Hypothesis Tests**.

4. Click the down arrow under Function type and select **1 Var *t* Test**.

5. Click **Hours** in the Names and Columns window. Then click the arrow under Quantitative
    Variable to select Hours for the analysis. Click **OK**.

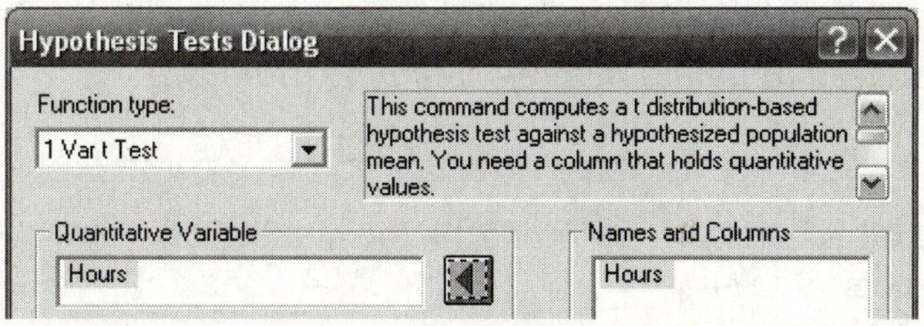

Hypothesis Tests Dialog

Function type:

1 Var t Test

This command computes a t distribution-based
hypothesis test against a hypothesized population
mean. You need a column that holds quantitative
values.

Quantitative Variable

Hours

Names and Columns

Hours

6. Click **Set $\mu 0$**. Enter **11** for the hypothesized population mean. Click **OK**.

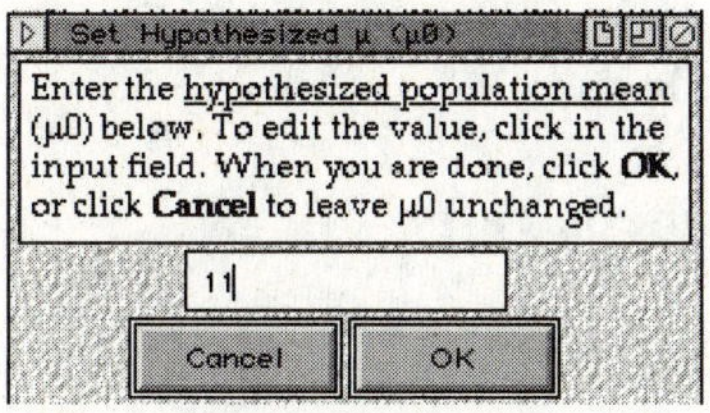

Set Hypothesized μ ($\mu 0$)

Enter the hypothesized population mean
($\mu 0$) below. To edit the value, click in the
input field. When you are done, click **OK**,
or click **Cancel** to leave $\mu 0$ unchanged.

11

Cancel   OK

7. Click **0.01** to set the significance level at 0.01.

8. Click $\mu \neq \mu 0$ to select a two-tailed test. Click **Compute**.

The output is shown below. The *t* statistic is -1.08 and the p-value is 0.3155.

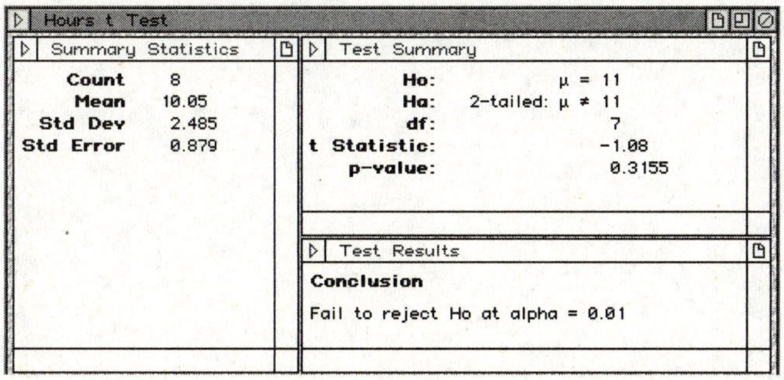

Hours t Test

Summary Statistics		Test Summary	
Count	8	Ho:	$\mu = 11$
Mean	10.05	Ha:	2-tailed: $\mu \neq 11$
Std Dev	2.485	df:	7
Std Error	0.879	t Statistic:	-1.08
		p-value:	0.3155

Test Results

Conclusion

Fail to reject Ho at alpha = 0.01

Section 7.4 Hypothesis Testing for Proportions

 **Example 1 (pg. 399)**   Hypothesis Test for a Proportion

If the DDXL add-in has not been loaded, you will need to load it before continuing. Follow the instructions in Section GS 8.2.

1. First open a new Excel worksheet. Type **39** in cell A1. This is 39% of 100, the number of successes. Type **100** in cell B1. This is the number of trials.

	A	B
1	39	100

2. Click and drag over the range **A1:B1** to select these numbers for the analysis.

3. At the top of the screen, click **Add-Ins** and select **DDXL**. Select **Hypothesis Tests**.

4. Click the down arrow under Function type and select **Summ 1 Var Prop Test**.

5. Select **$A$1** for Num Successes. Select **$B$1** for Num Trials. Click **OK**.

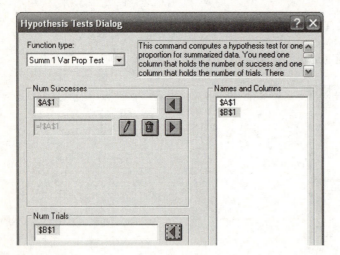

6. Click **Set μ0**. Enter **.50** for the hypothesized population proportion. Click **OK**.

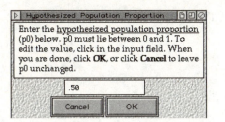

7. Click **.01** to set the significance level at 0.01.

8. For the alternative hypothesis, select *p < p0*. Click **Compute**.

The output is shown below. The *z* statistic is -2.2 and the *p*-value is 0.0139.

```
$A$1 Proportion Test

Summary Statistics          Test Summary
                                 p0:               0.5
     n      100                  Ho:          p = 0.5
  p-hat     0.39                 Ha:    Lower tail: p < 0.5
 Std Dev    0.05          z Statistic:              -2.2
                             p-value:             0.0139

                             Test Results
                          Conclusion
                          Fail to reject Ho at alpha =  0.01
```

◀

▶ Exercise 16 (pg. 402)	Testing the Claim That 30% of U.S. Households Own a Cat

If the DDXL add-in has not been loaded, you will need to load it before continuing. Follow the instructions in Section GS 8.2.

1. First open a new Excel worksheet. Type **72** in cell A1. This is the number of successes. Type **200** in cell B1. This is the number of trials.

2. Click and drag over the range **A1:B1** to select these numbers for the analysis.

3. At the top of the screen, click **Add-Ins** and select **DDXL**. Select **Hypothesis Tests**.

4.   Click the down arrow under Function type and select **Summ 1 Var Prop Test**.

5.   Select **$A$1** for Num Successes.   Select **$B$1** for Num Trials.   Click **OK**.

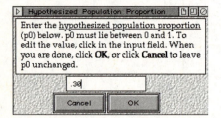

6.   Click **Set μ0**. Enter **.30** for the hypothesized population proportion. Click **OK**.

```
┌─ Hypothesized Population Proportion ──────┐
│ Enter the hypothesized population proportion │
│ (p0) below. p0 must lie between 0 and 1. To  │
│ edit the value, click in the input field. When│
│ you are done, click OK, or click Cancel to leave│
│ p0 unchanged.                                │
│                                              │
│     .30                                      │
│                                              │
│   ┌ Cancel ┐   ┌   OK   ┐                     │
└──────────────────────────────────────────┘
```

7.   Alpha has already been set at 0.05, so you do not need to select the significance level.   For the alternative hypothesis, select $p \neq p0$. Click **Compute**.

Your output should be the same as the output displayed below.   The *z* statistic is 1.85.   The *p*-value is 0.0641.

```
┌─ $A$1 Proportion Test ─────────────────────────────┐
│ ┌─ Summary Statistics ─┐ ┌─ Test Summary ─────────┐ │
│ │      n     200       │ │       p0:          0.3 │ │
│ │  p-hat    0.35       │ │       Ho:      p = 0.3 │ │
│ │ Std Dev   0.0324     │ │       Ha:  2-tailed: p ≠ 0.3 │
│ │                      │ │ z Statistic:      1.85 │ │
│ │                      │ │   p-value:      0.0641 │ │
│ │                      │ ├────────────────────────┤ │
│ │                      │ │ ┌─ Test Results ──────┐ │
│ │                      │ │ Conclusion            │ │
│ │                      │ │ Fail to reject Ho at alpha = 0.05 │
│ └──────────────────────┘ └────────────────────────┘ │
└────────────────────────────────────────────────────┘
```

Section 7.5 Hypothesis Testing for Variance and Standard Deviation

▶ Exercise 28 (pg. 412)	Testing the Claim That the Standard Deviation Is No More Than $30

If the DDXL add-in has not been loaded, you will need to load it before continuing. Follow the instructions in Section GS 8.2.

1. Open the "Ex7_5-28" worksheet in the Chapter 7 folder.

2. Click and drag over the data range, **A1:A21**, so that it is highlighted.

3. At the top of the screen, click **Add-Ins** and select **DDXL**. Select **Hypothesis Tests**.

4. Click the down arrow under Function type and select **Chisquare for SD**.

5. Click **Annual Salaries** in the Names and Columns window. Click the arrow below Quantitative Variable to select Annual Salaries for the analysis. Click **OK**.

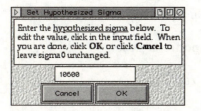

6. Click **Set Hypothesized Sigma**. Enter **10600**. Click **OK**.

7. Select **0.10** for the significance level.

8. Select **Left Tailed Test**. Click **Compute**.

Your output should be the same as the output displayed at the top of the next page. The chi-square statistic is 11.6. The *p*-value is 0.0983.

```
┌─────────────────────────────────────────────────────────┐
│ ▷  Annual Salaries Chi-square Test                ▣▣⊘    │
│ ┌───────────────────────────────────────────────────┐   │
│ │ ▷ │ Test Summary                                 ▣ │   │
│ │         Ho:              Sigma = 10600              │   │
│ │         Ha:   Lower tail: Sigma < 10600            │   │
│ │ Sample Std Dev                  8284.9             │   │
│ │  Test Statistic:                  11.6             │   │
│ │       P-value:                  0.0983             │   │
│ │                                                     │   │
│ └───────────────────────────────────────────────────┘   │
│ ┌───────────────────────────────────────────────────┐   │
│ │ ▷ │ Test Results                                 ▣ │   │
│ │ Conclusion                                          │   │
│ │ Reject Ho at alpha = 0.10                           │   │
│ │                                                     │   │
│ └───────────────────────────────────────────────────┘   │
└─────────────────────────────────────────────────────────┘
```

◄

Technology

▶ Exercise 1 (pg. 423)	Testing the Claim That the Proportion Is Equal to 0.53

If the DDXL add-in has not been loaded, you will need to load it before continuing. Follow the instructions in Section GS 8.2.

1. First open a new Excel worksheet. Type **102** in cell A1. This is the number of successes. Type **350** in cell B1. This is the number of trials.

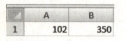

	A	B
1	102	350

Note that $0.2914 \times 350 = 102$. Note also that the Minitab illustration displays a 99.0% confidence interval. Therefore, the level of significance must be 0.01.

2. Click and drag over the range **A1:B1** to select these numbers for the analysis.

3. At the top of the screen, click **Add-Ins** and select **DDXL**. Select **Hypothesis Tests**.

4. Click the down arrow under Function type and select **Summ 1 Var Prop Test**.

5. Select **$A$1** for Num Successes. Select **$B$1** for Num Trials. Click **OK**.

There should not be a checkmark in the box next to First row is variable names.

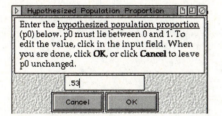

6. Click **Set μ0**. Enter **.53** for the hypothesized population proportion. Click **OK**.

7. Select **0.01** for the significance level.

8. Select $p \neq p0$ for a two-tailed test. Click **Compute**.

Your output should be the same as the output displayed below. The *z* statistic is -8.94. The *p*-value is < .0001.

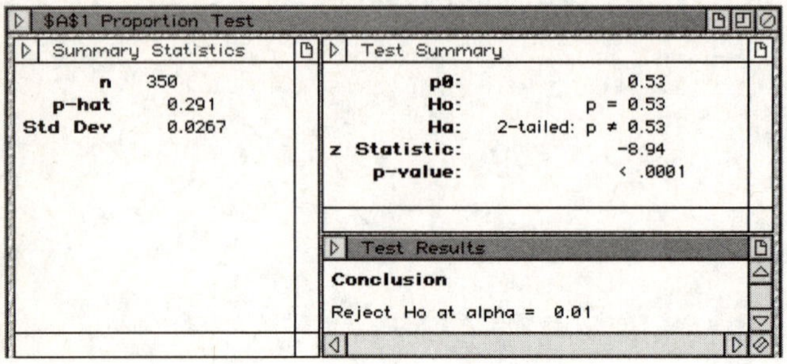

```
$A$1 Proportion Test

Summary Statistics          Test Summary

        n     350               p0:           0.53
    p-hat   0.291               Ho:       p = 0.53
  Std Dev  0.0267               Ha: 2-tailed: p ≠ 0.53
                        z Statistic:         -8.94
                            p-value:       < .0001


                        Test Results

                        Conclusion

                        Reject Ho at alpha =  0.01
```

Hypothesis Testing with Two Samples

Section 8.1 Testing the Difference Between Means (Large Independent Samples)

▶ Exercise 31 (pg. 437)	Testing the Claim That Children Ages 6-17 Watched TV More in 1981 Than Today

1. Open worksheet "Ex8_1-31" in the Chapter 8 folder.

2. Your textbook indicates that the variance of a sample may be used in the z-test formula instead of a known population variance if the sample size is sufficiently large. You will use Excel's Descriptive Statistics tool to obtain the sample variance. At the top of the screen, click **Data** and select **Data Analysis** in the Analysis group.

If Data Analysis does not appear as a choice in the Data ribbon, you will need to load the Microsoft Excel ToolPak add-in. Follow the procedure in Section GS 8.1 before continuing.

3. Select **Descriptive Statistics**. Click **OK**.

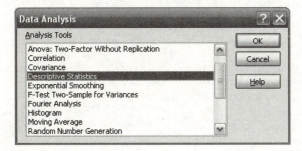

4.   Complete the Descriptive Statistics dialog box as shown below. Be sure to select **Labels in First Row** and **Summary statistics**. Click **OK**.

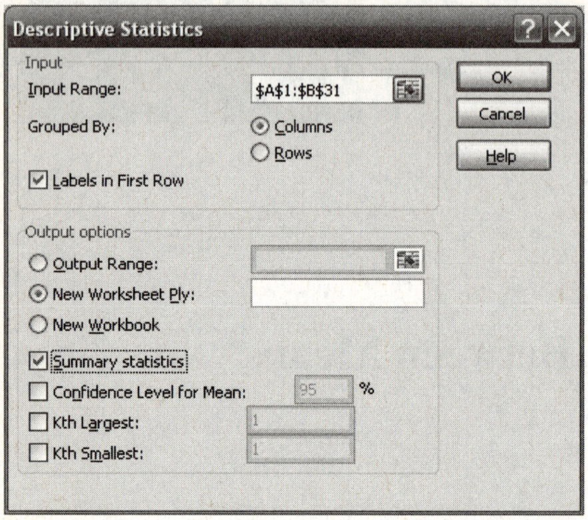

Descriptive Statistics   [?] [X]

Input
Input Range: $A$1:$B$31   [icon]   [OK]
Grouped By: ⦿ Columns   [Cancel]
                     ○ Rows   [Help]
☑ Labels in First Row

Output options
○ Output Range: [icon]
⦿ New Worksheet Ply: [   ]
○ New Workbook
☑ Summary statistics
☐ Confidence Level for Mean: 95   %
☐ Kth Largest: 1
☐ Kth Smallest: 1

5.   You will want to make column A and column C wider so that the labels are easier to read. The variance of Time A is 0.2401 and the variance of Time B is 0.2212. Return to the worksheet that contains the data. To do this, click on the **Sheet1** tab near the bottom of the screen.

	A	B	C	D
1	*Time A*		*Time B*	
2				
3	Mean	2.13	Mean	1.756667
4	Standard Error	0.089462	Standard Error	0.085861
5	Median	2.1	Median	1.7
6	Mode	2.1	Mode	1.6
7	Standard Deviation	0.490004	Standard Deviation	0.470277
8	Sample Variance	0.240103	Sample Variance	0.221161
9	Kurtosis	0.863836	Kurtosis	0.200666
10	Skewness	-0.00957	Skewness	0.531138
11	Range	2.3	Range	2
12	Minimum	1	Minimum	0.9
13	Maximum	3.3	Maximum	2.9
14	Sum	63.9	Sum	52.7
15	Count	30	Count	30

6. At the top of the screen, click **Data** and select **Data Analysis**. Select **z-Test: Two Sample for Means** and click **OK**.

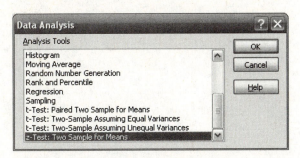

7. Complete the z-Test: Two Sample for Means dialog box as shown below. Click **OK**.

The output is displayed in a new worksheet. You will want to adjust the width of column A so that you can read the labels. Your output should look similar to the output displayed below.

	A	B	C
1	z-Test: Two Sample for Means		
2			
3		Time A	Time B
4	Mean	2.13	1.756667
5	Known Variance	0.2401	0.2212
6	Observations	30	30
7	Hypothesized Mean Difference	0	
8	z	3.010687	
9	P(Z<=z) one-tail	0.001303	
10	z Critical one-tail	1.959964	
11	P(Z<=z) two-tail	0.002607	
12	z Critical two-tail	2.241403	

◀

Section 8.2 Testing the Difference Between Means (Small Independent Samples)

▶ Exercise 19 (pg. 448)	Testing the Difference in Tensile Strength

If Data Analysis does not appear as a choice in the Data ribbon, you will need to load the Microsoft Excel ToolPak add-in. Follow the procedure in Section GS 8.1 before continuing.

1. Open worksheet "Ex8_2-19" in the Chapter 8 folder.

2. At the top of the screen, click **Data** and select **Data Analysis** in the Analysis group. Select **t-Test: Two-Sample Assuming Equal Variances**. Click **OK**.

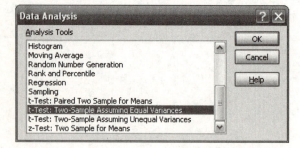

3. Complete the *t* Test: Two-Sample Assuming Equal Variances dialog box as shown below. Click **OK**.

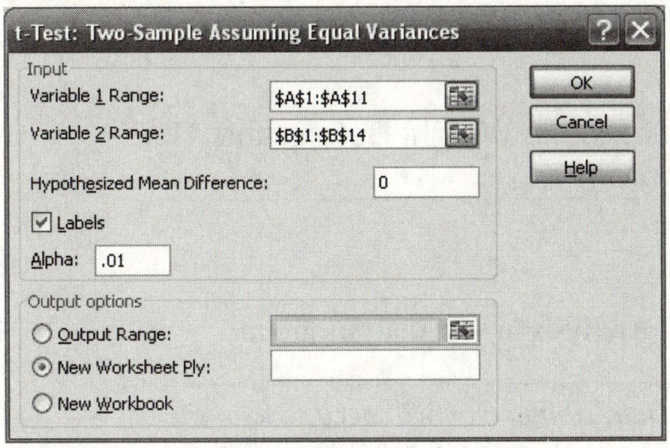

You will want to make the columns wider so that you can see all the output. The output is displayed below.

	A	B	C
1	t-Test: Two-Sample Assuming Equal Variances		
2			
3		New method	Old method
4	Mean	368.3	389.5384615
5	Variance	497.3444444	210.6025641
6	Observations	10	13
7	Pooled Variance	333.4919414	
8	Hypothesized Mean Difference	0	
9	df	21	
10	t Stat	-2.764955465	
11	P(T<=t) one-tail	0.005802501	
12	t Critical one-tail	2.517648014	
13	P(T<=t) two-tail	0.011605003	
14	t Critical two-tail	2.831359554	

Section 8.3 Testing the Difference Between Means (Dependent Samples)

▶ Exercise 11 (pg. 457)	Testing the Difference in Weight Before and After an Exercise Program

1. Open "Ex8_3-11 in the Chapter 8 folder.

2. At the top of the screen, click **Data** and select **Data Analysis** in the Analysis group.

If Data Analysis does not appear as a choice in the Data ribbon, you will need to load the Microsoft Excel ToolPak add-in. Follow the procedure in Section GS 8.1 before continuing.

3. Select *t*-**Test: Paired Two Sample for Means**. Click **OK**.

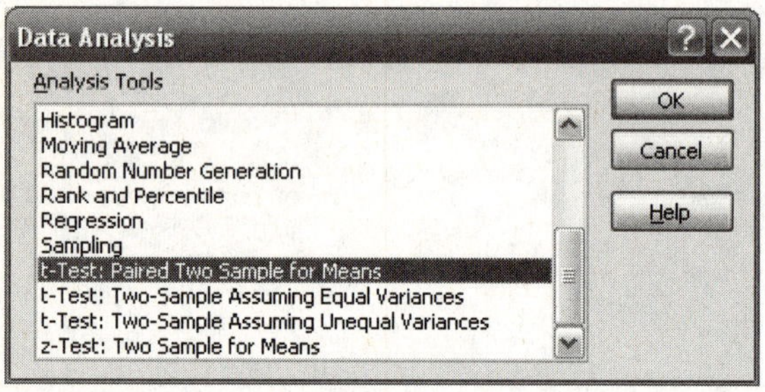

Data Analysis

Analysis Tools

- Histogram
- Moving Average
- Random Number Generation
- Rank and Percentile
- Regression
- Sampling
- t-Test: Paired Two Sample for Means
- t-Test: Two-Sample Assuming Equal Variances
- t-Test: Two-Sample Assuming Unequal Variances
- z-Test: Two Sample for Means

OK | Cancel | Help

4. Complete the *t*-Test: Paired Two Sample for Means dialog box as shown below. Be sure to select **Labels** and change **Alpha** to **.10**. Click **OK**.

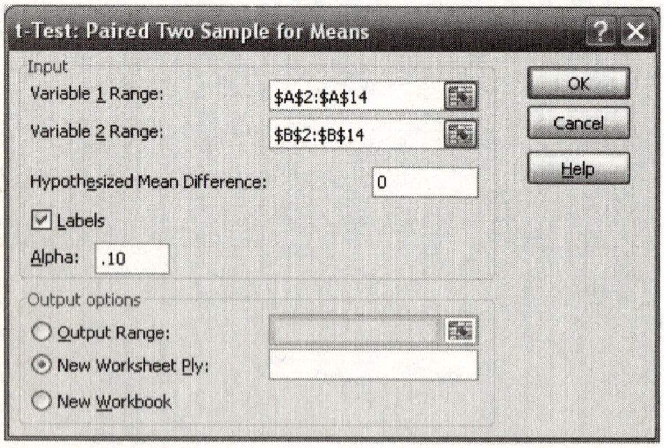

t-Test: Paired Two Sample for Means

Input
Variable 1 Range: $A$2:$A$14
Variable 2 Range: $B$2:$B$14

Hypothesized Mean Difference: 0

☑ Labels
Alpha: .10

Output options
○ Output Range:
◉ New Worksheet Ply:
○ New Workbook

OK | Cancel | Help

The output is displayed in a new worksheet. You will want to adjust the width of the columns so that you can read all the labels. Your output should look similar to the output displayed at the top of the next page.

	A	B	C
1	t-Test: Paired Two Sample for Means		
2			
3		*Before*	*After*
4	Mean	181	177.25
5	Variance	1584.545	1184.932
6	Observations	12	12
7	Pearson Correlation	0.988143	
8	Hypothesized Mean Difference	0	
9	df	11	
10	t Stat	1.656779	
11	P(T<=t) one-tail	0.06289	
12	t Critical one-tail	1.36343	
13	P(T<=t) two-tail	0.125781	
14	t Critical two-tail	1.795885	

◀

▶ Exercise 12 (pg. 457)	Testing the Difference in Batting Averages Before and After a Baseball Clinic

1.   Open worksheet "Ex8_3-12" in the Chapter 8 folder.

2.   At the top of the screen, click **Data** and select **Data Analysis** in the Analysis group.

If Data Analysis does not appear as a choice in the Data ribbon, you will need to load the Microsoft Excel ToolPak add-in. Follow the procedure in Section GS 8.1 before continuing.

---

3.   Select *t*-**Test: Paired Two Sample for Means**. Click **OK**.

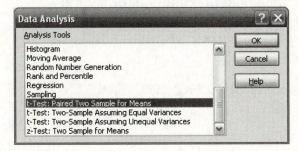

4. Complete the *t*-Test: Paired Two Sample for Means dialog box as shown below. Be sure to select **Labels**. Click **OK**.

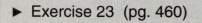

The output is displayed in a new worksheet. You will want to make the columns wider so that you can read all the labels. Your output should appear similar to the output shown below.

	A	B	C
1	t-Test: Paired Two Sample for Means		
2			
3		Before	After
4	Mean	0.308714	0.310429
5	Variance	0.001182	0.001008
6	Observations	14	14
7	Pearson Correlation	0.901664	
8	Hypothesized Mean Difference	0	
9	df	13	
10	t Stat	-0.43088	
11	P(T<=t) one-tail	0.336806	
12	t Critical one-tail	1.770933	
13	P(T<=t) two-tail	0.673613	
14	t Critical two-tail	2.160369	

◀

▶ Exercise 23 (pg. 460)	Constructing a 90% Confidence Interval for μ_D, the Mean Increase in Hours of Sleep

1. Open worksheet "Ex8_3-23" in the Chapter 8 folder.

2. Use Excel to calculate a difference score for each of the 16 patients. First, click in cell **C2** and key in the label **Difference**.

	A	B	C
1	Hours of sleep		
2	Without drug	With new drug	Difference
3	1.8	3.0	

3. Click in cell **C3** and enter the formula **=A3-B3** as shown below. Press [**Enter**].

	A	B	C
1		Hours of sleep	
2	Without drug	With new drug	Difference
3	1.8	3.0	=A3-B3

4. Click in cell **C3** (where −1.2 now appears) and copy the contents of that cell to cells C4 through C18.

	A	B	C
1		Hours of sleep	
2	Without drug	With new drug	Difference
3	1.8	3.0	-1.2
4	2.0	3.6	-1.6
5	3.4	4.0	-0.6
6	3.5	4.4	-0.9
7	3.7	4.5	-0.8
8	3.8	5.2	-1.4
9	3.9	5.5	-1.6
10	3.9	5.7	-1.8
11	4.0	6.2	-2.2
12	4.9	6.3	-1.4
13	5.1	6.6	-1.5
14	5.2	7.8	-2.6
15	5.0	7.2	-2.2
16	4.5	6.5	-2.0
17	4.2	5.6	-1.4
18	4.7	5.9	-1.2

5. You will now obtain descriptive statistics for Difference. At the top of the screen, click **Data** and select **Data Analysis** in the Analysis group.   Select **Descriptive Statistics**. Click **OK**.

If Data Analysis does not appear as a choice in the Data ribbon, you will need to load the Microsoft Excel ToolPak add-in. Follow the procedure in Section GS 8.1 before continuing.

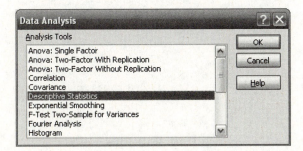

6. Complete the Descriptive Statistics dialog box as shown below. Be sure to select Summary statistics and change the Confidence Level for Mean to 90%. Click **OK**.

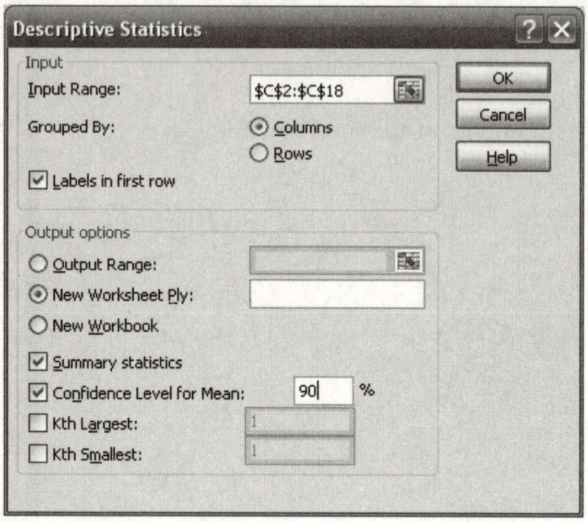

Descriptive Statistics	? X
Input	
Input Range:	$C$2:$C$18
Grouped By:	⦿ Columns
	○ Rows
☑ Labels in first row	
	OK
	Cancel
	Help
Output options	
○ Output Range:	
⦿ New Worksheet Ply:	
○ New Workbook	
☑ Summary statistics	
☑ Confidence Level for Mean:	90 %
☐ Kth Largest:	1
☐ Kth Smallest:	1

The output is displayed in a new worksheet. The mean is printed at the top and the confidence level is printed at the bottom. To calculate the lower limit of the 90% confidence interval, subtract 0.2376 from –1.525. To calculate the upper limit, add 0.2376 to –1.525.

	A	B
1	*Difference*	
2		
3	Mean	-1.525
4	Standard Error	0.135554
5	Median	-1.45
6	Mode	-1.6
7	Standard Deviation	0.542218
8	Sample Variance	0.294
9	Kurtosis	-0.26069
10	Skewness	-0.23658
11	Range	2
12	Minimum	-2.6
13	Maximum	-0.6
14	Sum	-24.4
15	Count	16
16	Confidence Level(90.0%)	0.237634

Section 8.4 Testing the Difference Between Proportions

▶ Example 1 (pg. 463) | A Two-Sample *z*-Test for the Difference Between Proportions

If the DDXL add-in has not been loaded, you will need to load it before continuing. Follow the instructions in Section GS 8.2.

1. First open a new Excel worksheet. Then, enter the number of successes and number of trials for passenger cars (129 and 150) and for pickup trucks (148 and 200) in row 1 of the worksheet as shown below.

	A	B	C	D
1	129	150	148	200

2. Click and drag over the data range, **A1:D1**, to select these numbers for the analysis.

3. At the top of the screen, click **Add-Ins** and select **DDXL**. Select **Hypothesis Tests**.

4. Under Function type, select **Summ 2 Var Prop Test**.

5. Select the appropriate cell address for each of the four windows on the left as shown below. Click **OK**.

Note that there is no checkmark in the box next to First row is variable names.

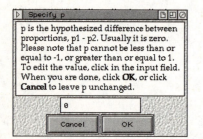

Hypothesis Tests Dialog

Function type:
Summ 2 Var Prop Test

This command computes a hypothesis test for the difference between two proportions for summarized data. You need one column that holds the number of successes for the first

Num Successes 1
$A$1
=!$A$1

Num Trials 1
$B$1
=!$B$1

Num Successes 2
$C$1
=!$C$1

Num Trials 2
$D$1
=!$D$1

Names and Columns
$A$1
$B$1
$C$1
$D$1

☐ First row is variable names

Info   Help

Cancel   OK

6. Set the hypothesized difference equal to 0 as shown below. Click **OK**.

Specify p

p is the hypothesized difference between proportions, p1 - p2. Usually it is zero. Please note that p cannot be less than or equal to -1, or greater than or equal to 1. To edit the value, click in the input field. When you are done, click **OK**, or click **Cancel** to leave p unchanged.

0

Cancel   OK

7. Click **.10** to set the significance level equal to 0.10.

8. Click $p1 - p2 \neq 0$ for a two-tailed test. Click **Compute**.

The output is shown below. The *z* statistic is 2.73. The *p*-value is 0.0062.

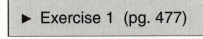

```
┌─────────────────────────────────────────────────────────────────────────────┐
│ ▷ Proportion Test for the Difference Between $A$1 and $B$1          ▯ ▯ ⊘     │
├─────────────────────────────┬─────────────────────────────────────────────────┤
│ ▷  Summary Statistics    ▯  │ ▯ ▷  Test Summary                            ▯  │
│                             │                                                 │
│       n1      150           │              p:                  0              │
│    p-hat1     0.86          │             Ho:          p1 - p2 = 0            │
│       n2      200           │             Ha:   2-tailed: p1 - p2 ≠ 0         │
│    p-hat2     0.74          │   z Statistic:                2.73              │
│  Difference   0.12          │      p-value:                0.0062             │
│    pooled n   350           │                                                 │
│ pooled p-hat  0.791         ├─────────────────────────────────────────────┐  │
│    Std Err    0.0439        │ ▷  Test Results                          ▯  │  │
│                             │ Conclusion                               △  │  │
│                             │                                             │  │
│                             │ Reject Ho at alpha =  0.10               ▽  │  │
│                             │ ◁                                    ▷ ◈    │  │
└─────────────────────────────┴─────────────────────────────────────────────────┘
```

◀

Technology

▶ Exercise 1 (pg. 477)	Testing the Hypothesis That the Probability of a "Found Coin" Lying Heads Up Is 0.5

If the DDXL add-in has not been loaded, you will need to load it before continuing. Follow the instructions in Section GS 8.2.

1. First open a new Excel worksheet. Then, enter the number of coins found heads up (5772) and the total number of coins found (11902) in cells A1 and B1 as shown below.

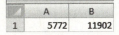

	A	B
1	5772	11902

2. Click and drag over the data range **A1:B1** to select these numbers for the analysis.

3. At the top of the screen, click **Add-Ins** and select **DDXL**. Select **Hypothesis Tests**.

4.   Under Function type, select **Summ 1 Var Prop Test**. Select cell $A$1 for number of successes and select cell $B$1 for number of trials. Click **OK**.

Note that there is no checkmark in the box next to First row is variable names.

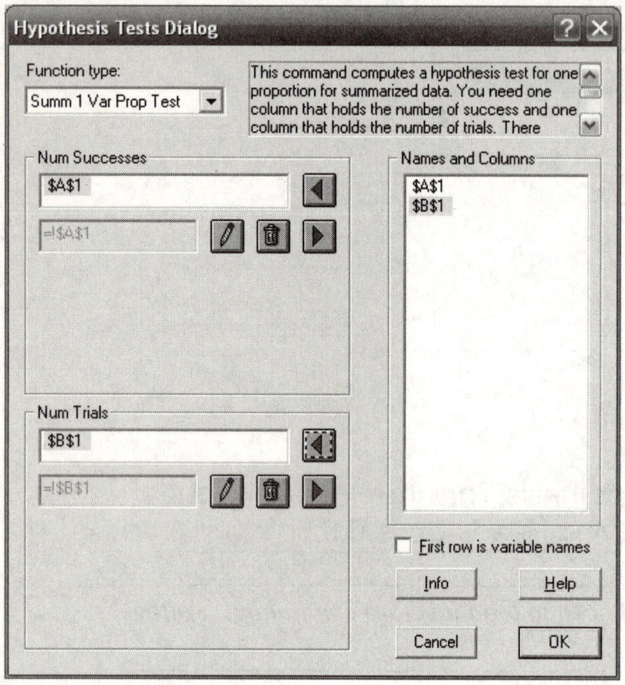

5.   Click **Set $p0$**. Enter **.50** for the hypothesized population proportion. Click **OK**.

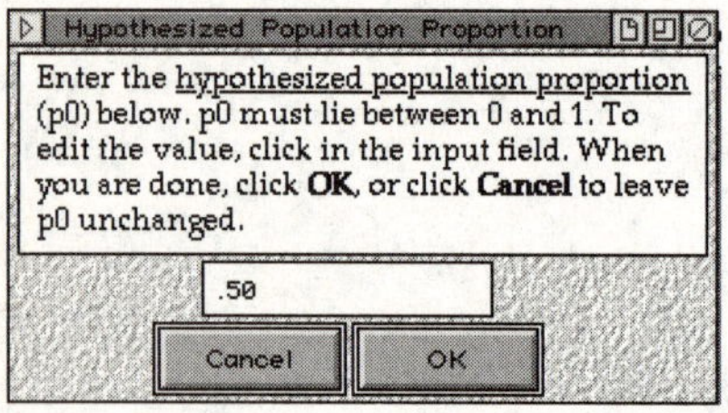

6.   Click **.01** to set the significance level at 0.01.

7.   Select $p \neq p0$ for a two-tailed test. Click **Compute**.

The output is shown below. The *z* statistic is -3.28 and the *p*-value is 0.001.

```
┌▷┬ $A$1 Proportion Test ──────────────────────────────┬□□∅┐
│┌▷┬ Summary Statistics ─────┬□┐┌▷┬ Test Summary ────────────────┬□┐│
││                          ││││          p0:        0.5          ││
││     n    11902           ││││          Ho:    p = 0.5          ││
││  p-hat      0.485        ││││          Ha:  2-tailed: p ≠ 0.5  ││
││ Std Dev     0.00458      ││││   z Statistic:      -3.28        ││
││                          ││││      p-value:        0.001       ││
││                          ││││                                  ││
││                          ││││                                  ││
││                          ││││                                  ││
││                          ││││┌▷┬ Test Results ──────────────┬□┐││
││                          ││││ Conclusion                     │││
││                          ││││ Reject Ho at alpha =  0.01     │││
│└──────────────────────────┘│└──────────────────────────────┘ ││
└────────────────────────────┴──────────────────────────────────┘
```

◀

► **Exercise 3 (pg. 477)**   Simulating "Tails Over Heads"

---

1.   Open a new Excel worksheet.

2.   At the top of the screen, click **Data** and select **Data Analysis** in the Analysis group.

If Data Analysis does not appear in the Data ribbon, you will need to load the Microsoft Excel ToolPak add-in. Follow the procedure in Section GS 8.1 before continuing.

---

3.   Select **Random Number Generation**. Click **OK**.

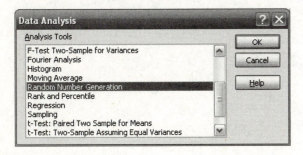

4.   Complete the Random Number Generation dialog box as shown below. The output will be placed in cell A1. Click **OK**.

Random Number Generation   ? ✕

Number of Variables:	1	OK
Number of Random Numbers:	1	Cancel
Distribution:	Binomial ▾	Help

Parameters

p Value = 0.5

Number of trials = 11902

Random Seed:

Output options
⦿ Output Range: $A$1
◯ New Worksheet Ply:
◯ New Workbook

The output for this simulation indicates that 5,934 of the 11,902 coins were found heads up. Because this output was generated randomly, it is unlikely that your output will be exactly the same.

	A
1	5934

◀

Correlation and Regression

Section 9.1 Correlation

► Example 3 (pg. 486)   Constructing a Scatter Plot Using Technology

1. Open worksheet "Old Faithful" in the Chapter 9 folder.

2. Click on any cell within the table of data. At the top of the screen, click **Insert** and select **Scatter** from the Charts group.

3. In the scatter type box, select the leftmost chart in the top row by clicking on it.

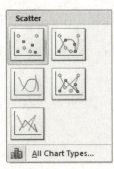

4. In the Design ribbon, click the leftmost diagram (**Layout 1**) in Chart Layouts so that you can add an X-axis title, a Y-axis title, and revise the main chart title.

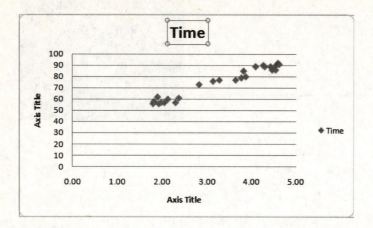

5. Click on **Time** and replace it with **Duration of Old Faithful's Eruptions by Time Until the Next Eruption**. Click on the Y **Axis Title** and replace it with **Time Until the Next Eruption (in minutes)**. Click on the X **Axis Title** and replace it with **Duration (in minutes)**.

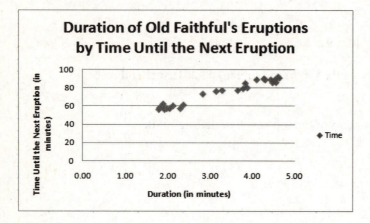

6. To remove the legend on the right, click on **Time** and press [**Delete**].

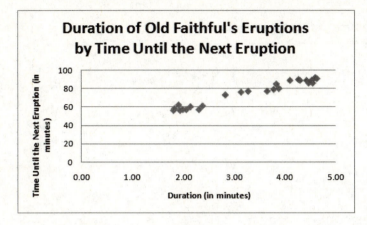

7. Let's move the scatter plot to a new worksheet. To do this, **right-click** in the white area near a border and select **Move Chart** from the menu that appears.

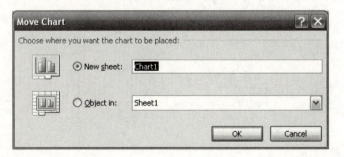

8. Select **New sheet** and click **OK**:

Your scatter plot should look similar to the one shown below.

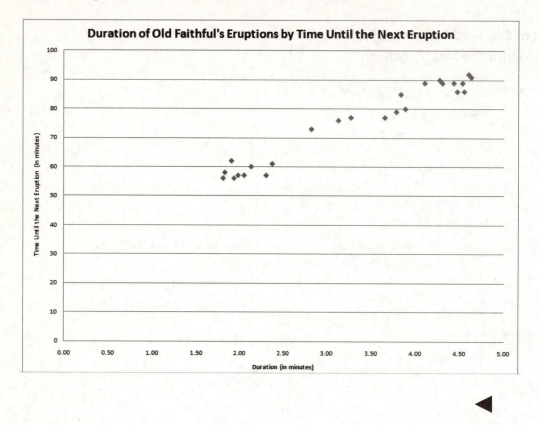

▶ Example 5 (pg. 489)   Using Technology to Find a Correlation Coefficient

1. Open worksheet "Old Faithful" in the Chapter 9 folder and click in cell **C1** to place the output there.

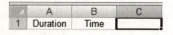

2. At the top of the screen, select **Formulas** and **Insert Function**.

3. Select the **Statistical** category. Select the **CORREL** function. Click **OK**.

Insert Function

Search for a function:

Type a brief description of what you want to do and then click Go   [Go]

Or select a _c_ategory: Statistical

Select a functio_n_:

CHIINV
CHITEST
CONFIDENCE
CORREL
COUNT
COUNTA
COUNTBLANK

CORREL(array1,array2)
Returns the correlation coefficient between two data sets.

Help on this function   [OK]   [Cancel]

4. Complete the CORREL dialog box as shown below. Click **OK**.

Function Arguments

CORREL

Array1 A1:A26   = {"Duration";1.8;1.82;1.9;1.93;1.98;...

Array2 B1:B26   = {"Time";56;58;62;56;57;57;60;57;61...

= 0.978659213

Returns the correlation coefficient between two data sets.

Array2 is a second cell range of values. The values should be numbers, names, arrays, or references that contain numbers.

Formula result = 0.978659213

Help on this function   [OK]   [Cancel]

Your output should look like the output shown below.

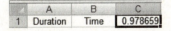

	A	B	C
1	Duration	Time	0.978659

◀

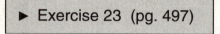 ► **Exercise 23 (pg. 497)**   Scatter Plot and Correlation for Hours Studying and Test Scores

1.   Open worksheet "Ex9_1-23" in the Chapter 9 folder.

2.   Click on any cell within the table of data. At the top of the screen, click **Insert** and select **Scatter** in the Charts group.

3.   In the scatter type box, select the leftmost chart in the top row by clicking on it.

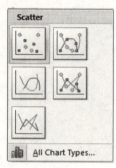

4.   In the Design ribbon, click the leftmost diagram (**Layout 1**) from Chart Layouts so that you can add an X-axis title, a Y-axis title, and revise the main chart title.

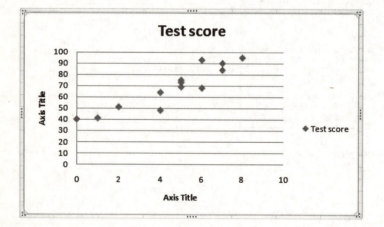

5. Click on **Test Score** and replace it with **Hours Spent Studying for a Test by Test Score**. Click on the Y **Axis Title** and replace it with **Test Score**. Click on the X **Axis Title** and replace it with **Hours**.

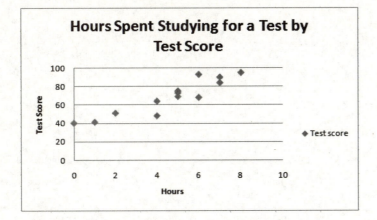

6. To remove the legend on the right, click on **Test score** and press [**Delete**].

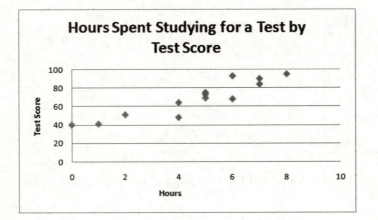

7. Let's move the scatter plot to a new worksheet. To do this, **right-click** in the white area near a border and select **Move Chart** from the menu that appears.

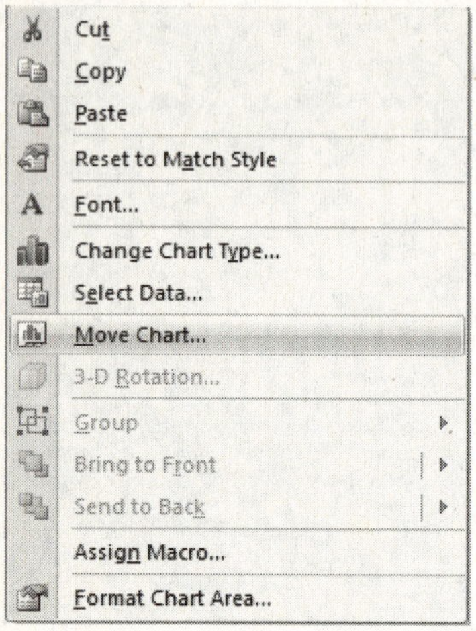

✂	Cu̲t
🖺	C̲opy
🖺	Paste
🖼	Reset to M̲atch Style
A	F̲ont...
ᵭ	Change Chart T̲ype...
🗒	S̲elect Data...
📊	Move Chart...
	3-D R̲otation...
🖽	Group ▸
🖺	Bring to F̲ront ▸
🖺	Send to Ba̲ck ▸
	Assi̲gn Macro...
🖼	F̲ormat Chart Area...

8. Select **New sheet** and click **OK**.

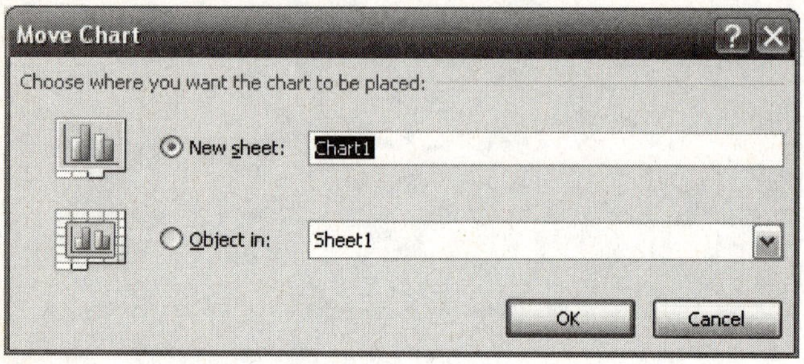

Move Chart [?][X]

Choose where you want the chart to be placed:

⊙ New s̲heet: `Chart1`

○ O̲bject in: `Sheet1` ▾

[OK] [Cancel]

Your scatter plot should look similar to the one shown at the top of the next page.

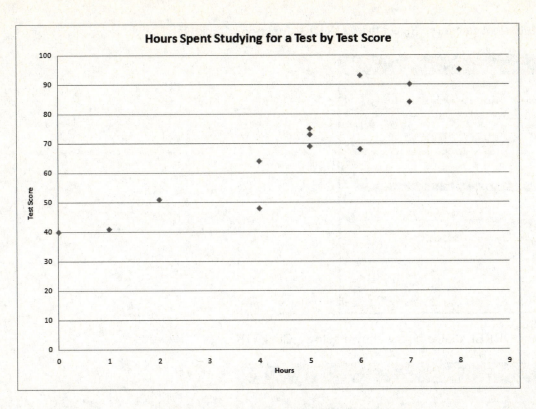

9. You will now use Excel to find the correlation between hours and test score. Return to the worksheet containing the data by clicking the **Sheet1** tab near the bottom of the screen. Click in cell **C1** of the worksheet to place the output there.

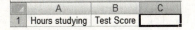

10. At the top of the screen, select **Formulas** and **Insert Function**.

11. Select the **Statistical** category. Select the **CORREL** function. Click **OK**.

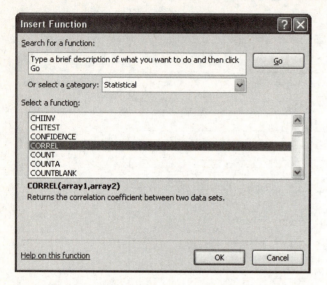

12. Complete the CORREL dialog box as shown below. Click **OK**.

Function Arguments

CORREL

Array1 A1:A14 = {"Hours studying";0;1;2;4;4;5;5;5;6;...

Array2 B1:B14 = {"Test score";40;41;51;48;64;69;73;...

= 0.922649866

Returns the correlation coefficient between two data sets.

Array2 is a second cell range of values. The values should be numbers, names, arrays, or references that contain numbers.

Formula result = 0.922649866

Help on this function   OK   Cancel

Your output should look like the output shown below.

	A	B	C
1	Hours studying	Test Score	0.92265
2	0	40	

Section 9.2 Linear Regression

▶ **Example 2 (pg. 503)**   Using Technology to Find a Regression Equation

1. Open worksheet "Old Faithful" in the Chapter 9 folder.

2. You will be using Excel's functions to obtain the slope and intercept. Type the labels **Slope** and **Intercept** in the worksheet as shown below. Then click in cell **D1** where the slope will be placed.

	A	B	C	D
1	Duration	Time	Slope	
2	1.80	56	Intercept	

3. At the top of the screen, select **Formulas** and **Insert Function**.

4. Select the **Statistical** category. Select the **SLOPE** function. Click **OK**.

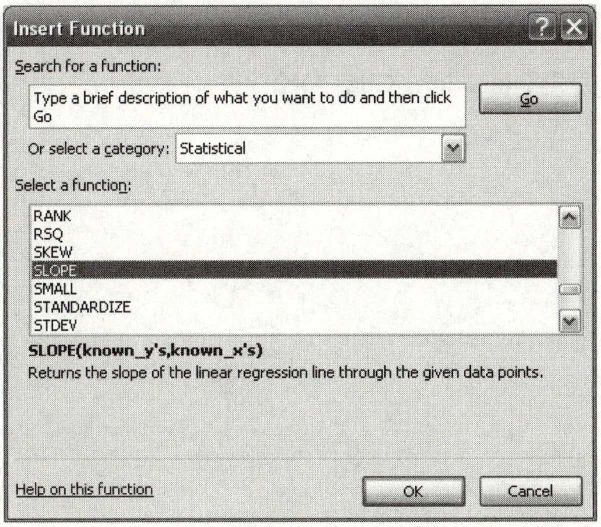

5. Complete the SLOPE dialog box as shown below. Click **OK**.

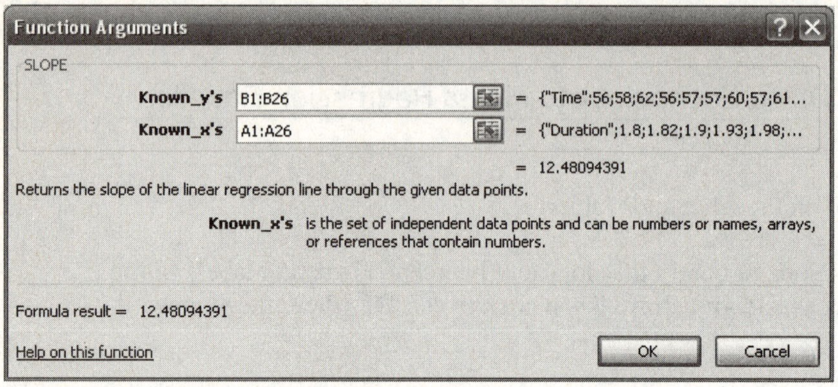

Function Arguments   ? ✕

SLOPE

       Known_y's   B1:B26   ▦   = {"Time";56;58;62;56;57;57;60;57;61...

       Known_x's   A1:A26   ▦   = {"Duration";1.8;1.82;1.9;1.93;1.98;...

                                                      = 12.48094391

Returns the slope of the linear regression line through the given data points.

                  Known_x's is the set of independent data points and can be numbers or names, arrays,
                               or references that contain numbers.

Formula result = 12.48094391

Help on this function   OK   Cancel

6. A slope of 12.48 is returned and placed in cell D1 of the worksheet. Click in cell **D2** where the intercept will be placed.

	A	B	C	D
1	Duration	Time	Slope	12.48094
2	1.80	56	Intercept	

7. At the top of the screen, select **Formulas** and **Insert Function**.

8. Select the **Statistical** category. Select the **INTERCEPT** function. Click **OK**.

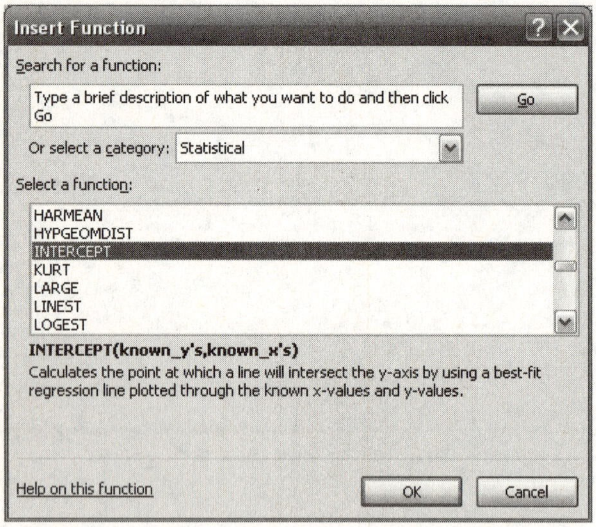

Insert Function   ? ✕

Search for a function:

   Type a brief description of what you want to do and then click   Go
   Go

Or select a category: Statistical   ▾

Select a function:

   HARMEAN
   HYPGEOMDIST
   INTERCEPT
   KURT
   LARGE
   LINEST
   LOGEST

INTERCEPT(known_y's,known_x's)
Calculates the point at which a line will intersect the y-axis by using a best-fit
regression line plotted through the known x-values and y-values.

Help on this function   OK   Cancel

9.   Complete the INTERCEPT dialog box as shown below. Click **OK**.

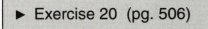

Function Arguments	? X

INTERCEPT

Known_y's   B1:B26   = {"Time";56;58;62;56;57;57;60;57;61...

Known_x's   A1:A26   = {"Duration";1.8;1.82;1.9;1.93;1.98;...

= 33.68290034

Calculates the point at which a line will intersect the y-axis by using a best-fit regression line plotted through the known x-values and y-values.

Known_x's   is the independent set of observations or data and can be numbers or names, arrays, or references that contain numbers.

Formula result =   33.68290034

Help on this function   OK   Cancel

An intercept of 33.68 is returned and placed in cell D2 of the worksheet.

	A	B	C	D
1	Duration	Time	Slope	12.48094
2	1.80	56	Intercept	33.6829

◄

▶ Exercise 20 (pg. 506)   **Finding the Equation of a Regression Line for Hours Online and Test Scores**

1.   Open worksheet "Ex9_2-20" in the Chapter 9 folder.

2.   At the top of the screen, click **Data** and select **Data Analysis** in the Analysis group. Select **Regression** and click **OK**.

If Data Analysis does not appear as a choice in the Data ribbon, you will need to load the Microsoft Excel Analysis ToolPak add-in. Follow the procedure in Section GS 8.1 before continuing.

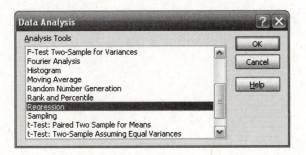

Data Analysis	? X

Analysis Tools

F-Test Two-Sample for Variances   OK
Fourier Analysis
Histogram   Cancel
Moving Average
Random Number Generation   Help
Rank and Percentile
Regression
Sampling
t-Test: Paired Two Sample for Means
t-Test: Two-Sample Assuming Equal Variances

3. Complete the Regression dialog box as shown below. Click **OK**.

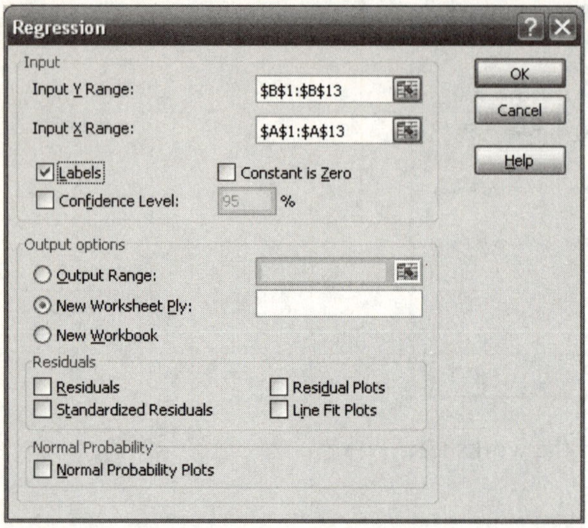

The intercept and slope of the regression equation are shown in the bottom two lines of the output under the label "Coefficients." The intercept is 93.97 and the slope is –4.07.

16		Coefficients	Standard Error	t Stat	P-value	Lower 95%	Upper 95%
17	Intercept	93.97003745	4.523586141	20.77335	1.48E-09	83.890859	104.049215
18	Hours onlin	-4.06741573	0.860011871	-4.72949	0.000805	-5.983642	-2.1511899

◄

Section 9.3 Measures of Regression and Prediction Intervals

► Example 2 (pg. 516)   Finding the Standard Error of Estimate

1. Open a new Excel worksheet and enter the expenses and sales data as shown at the top of the next page.

	A	B
1	Gross Domestic Product	Carbon Dioxide Emissions
2	1.6	428.2
3	3.6	828.8
4	4.9	1214.2
5	1.1	444.6
6	0.9	264
7	2.9	415.3
8	2.7	571.8
9	2.3	454.9
10	1.6	358.7
11	1.5	573.5

2.   At the top of the screen, click **Data** and select **Data Analysis** in the Analysis group. Select **Regression** and click **OK**.

If Data Analysis does not appear as a choice in the Data ribbon, you will need to load the Microsoft Excel Analysis ToolPak add-in. Follow the procedure in Section GS 8.1 before continuing.

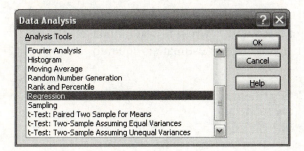

3.   Complete the Regression dialog box as shown below. If you select **Residuals**, the output will include $y - \hat{y}$ for each observation in the data set. Click **OK**.

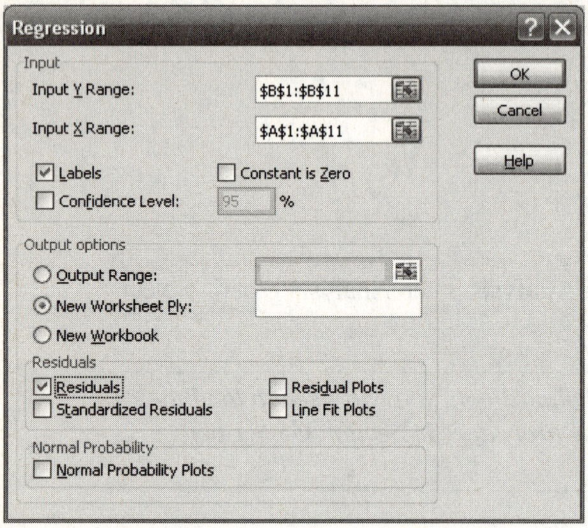

The standard error of estimate is displayed in the top part of the output. You will want to make column A wider so that you can read all the labels. The standard error of estimate is equal to 138.255.

	A	B
1	SUMMARY OUTPUT	
2		
3	*Regression Statistics*	
4	Multiple R	0.882469
5	R Square	0.778752
6	Adjusted R Square	0.751096
7	Standard Error	138.2552
8	Observations	10

The residuals are displayed in the lower part of the output.

22	RESIDUAL OUTPUT		
23			
24	*Observation*	*Predicted Carbon Dioxide Emissions*	*Residuals*
25	1	416.1320966	12.06790335
26	2	808.4360498	20.36395025
27	3	1063.433619	150.7663807
28	4	318.0561084	126.5438916
29	5	278.8257131	-14.82571306
30	6	671.1296662	-255.8296662
31	7	631.8992709	-60.09927086
32	8	553.4384802	-98.53848023
33	9	416.1320966	-57.43209665
34	10	396.516899	176.983101

◀

Section 9.4 Multiple Regression

▶ Example 1 (pg. 524)   Finding a Multiple Regression Equation

1. Open worksheet "Salary" in the Chapter 9 folder.

2. At the top of the screen, click **Data** and select **Data Analysis** in the Analysis group. Select **Regression** and click **OK**.

If Data Analysis does not appear as a choice in the Data ribbon, you will need to load the Microsoft Excel Analysis ToolPak add-in. Follow the procedure in Section GS 8.1 before continuing.

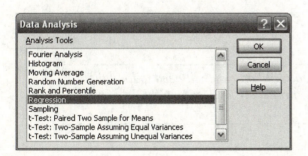

3. Complete the Regression dialog box as shown at the top of the next page. Click **OK**.

The coefficients for the multiple regression equation are displayed in the lower portion of the output under the label "Coefficients."

16		Coefficients	Standard Error	t Stat	P-value	Lower 95%
17	Intercept	49764.446	1981.346465	25.11648	1.49E-05	44263.35
18	Employment (yrs)	364.4120281	48.31750816	7.542029	0.001656	230.2611
19	Experience (yrs)	227.6188106	123.8361513	1.838064	0.139912	-116.2055
20	Education (yrs)	266.9350412	147.3556227	1.811502	0.144295	-142.1898

◄

▶ Exercise 5 (pg. 528)   Finding a Multiple Regression Equation, the
                              Standard Error of Estimate, and R^2

1. Open worksheet "Ex9_4-5" in the Chapter 9 folder.

2. At the top of the screen, click **Data** and select **Data Analysis** in the Analysis group. Select **Regression** and click **OK**.

If Data Analysis does not appear as a choice in the Data ribbon, you will need to load the Microsoft Excel Analysis ToolPak add-in. Follow the procedure in Section GS 8.1 before continuing.

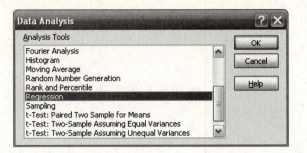

3.   Complete the Regression dialog box as shown below. Click **OK**.

The output is shown at the top of the next page. R^2 is equal to 0.9850. The standard error of estimate is equal to 28.489. The coefficients for the multiple regression equation are displayed in the lower portion of the output under the label "Coefficients."

	A	B
1	SUMMARY OUTPUT	
2		
3	*Regression Statistics*	
4	Multiple R	0.992469
5	R Square	0.984995
6	Adjusted R Square	0.981243
7	Standard Error	28.489
8	Observations	11

16		Coefficients
17	Intercept	-2518.36355
18	Square footage	126.8217972
19	Shopping centers	66.35999407

◄

Technology

▶ Exercise 1 (pg. 537) Constructing a Scatter Plot

Directions are given here for constructing a scatter plot of the calories and sugar variables. You can follow these instructions for constructing scatter plots of the other pairs of variables, but you will want to copy the pairs of variables to be graphed to another location in the worksheet so that they are located in adjacent columns. The x variable should be placed to the left of the y variable.

1. Open worksheet "Tech9" in the Chapter 9 folder.

2. You are asked to construct a scatter plot of calories (C) and sugar (S) where calories is the x variable and sugar is the y variable. Copy and paste the data in columns G and H of the worksheet so that the calories variable (x) is placed to the left of the sugar variable (y). The first few lines are shown below.

G	H
C	S
100	12
130	11
100	1

3. Click and drag over the calories and sugar data in columns G and H so that the range **G1:H22** is highlighted. At the top of the screen, click **Insert** and select **Scatter** in the Charts group.

4. In the scatter type box, select the leftmost chart in the top row by clicking on it.

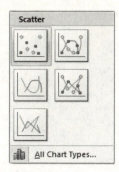

5. To add a trendline and to display the equation and R^2, **right-click** directly on a point in the plot.

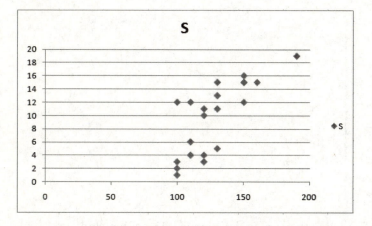

6. Select **Add Trendline** from the shortcut menu that appears.

7.   Select a **Linear** trend/regression. At the bottom of the dialog box, click in the boxes to display the equation and R-squared. Click **Close**.

```
Format Trendline                                              ? X

  Trendline Options      Trendline Options
  Line Color               Trend/Regression Type
  Line Style                  [icon]   ○  Exponential
  Shadow
                            [icon]   ⊙  Linear

                            [icon]   ○  Logarithmic

                            [icon]   ○  Polynomial     Order:  2

                            [icon]   ○  Power

                            [icon]   ○  Moving Average   Period:  2

                           Trendline Name
                             ⊙  Automatic :     Linear (S)
                             ○  Custom:       [            ]

                           Forecast
                             Forward:   0.0            periods
                             Backward:  0.0            periods

                           ☐  Set Intercept =   0.0
                           ☑  Display Equation on chart
                           ☑  Display R-squared value on chart

                                                    [  Close  ]
```

8.   You may want to move the equation and R^2 so that they are easier to read. The completed scatter plot is shown at the top of the next page.

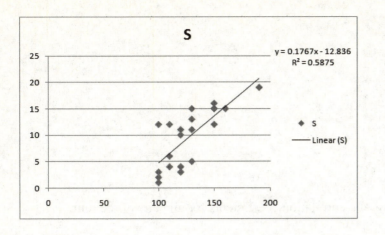

◄

| ► Exercise 3 (pg. 537) | Find the Correlation Coefficient for Each Pair of Variables |

1. Open worksheet "Tech9" in the Chapter 9 folder.

2. You will be using Excel to construct a correlation matrix with all four quantitative variables in the data set. At the top of the screen, click **Data** and select **Data Analysis** in the Analysis group. Select **Correlation** and click **OK**.

If Data Analysis does not appear as a choice in the Data ribbon, you will need to load the Microsoft Excel Analysis ToolPak add-in. Follow the procedure in Section GS 8.1 before continuing.

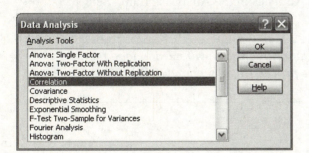

3. Complete the Correlation dialog box as shown at the top of the next page. Be sure to select **Labels in First Row**. Click **OK**.

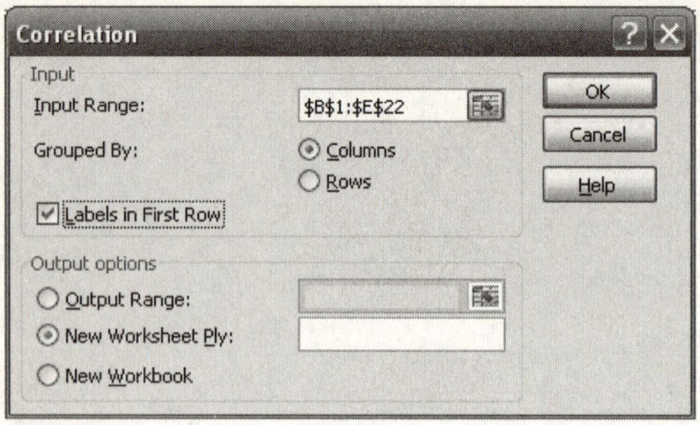

The output is a correlation matrix that displays the correlation coefficients for all pairs of the four variables.

	A	B	C	D	E
1		C	S	F	R
2	C	1			
3	S	0.766456	1		
4	F	0.415035	0.461382	1	
5	R	0.912751	0.793239	0.230294	1

◀

► **Exercise 4 (pg. 537)**   Finding the Equation of a Regression Line

---

Directions are given here for finding the equation for calories (X) and sugar (Y).

1. Open worksheet "Tech9" in the Chapter 9 folder.

2. At the top of the screen, click **Data** and select **Data Analysis** in the Analysis group. Select **Regression** and click **OK**.

If Data Analysis does not appear as a choice in the Data ribbon, you will need to load the Microsoft Excel Analysis ToolPak add-in. Follow the procedure in Section GS 8.1 before continuing.

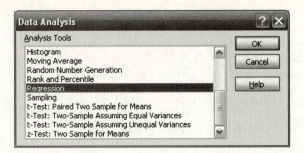

3. Complete the Regression dialog box as shown below. Click **OK**.

The coefficients for the analysis are given in the lower part of the output. The intercept is -12.836 and the slope is 0.177.

16		Coefficients
17	Intercept	-12.83645656
18	C	0.176703578

◀

▶ Exercise 6 (pg. 537)	Finding Multiple Regression Equations

Directions are given here for Exercise 6 (a) $C = b + m_1S + m_2F + m_3R$.

1.   Open worksheet "Tech9" in the Chapter 9 folder.

*If you have just completed Exercise 1, Exercise 3, or Exercise 4 on page 537 and have not closed the Excel worksheet, return to the sheet containing the data by clicking on the **Sheet1** tab at the bottom of the screen.*

2.   At the top of the screen, click **Data** and select **Data Analysis** in the Analysis group. Select **Regression** and click **OK**.

If Data Analysis does not appear as a choice in the Data ribbon, you will need to load the Microsoft Excel Analysis ToolPak add-in. Follow the procedure in Section GS 8.1 before continuing.

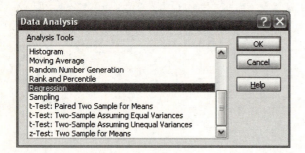

3.   Complete the Regression dialog box as shown below. Click **OK**.

The predictor variables must be located in adjacent columns in the Excel worksheet.

The coefficients for the equation are given in the lower part of the output.

16		Coefficients
17	Intercept	12.92908032
18	S	-0.268276766
19	F	7.567206446
20	R	3.88567347

Chi-Square Tests and the F-Distribution

Section 10.2 Independence

> ▶ Example 2 (pg.554)   Performing a Chi-Square Independence Test

---

In order to use Excel to carry out a chi-square test of independence, you must have already calculated the expected frequencies. Example 2 provides both the observed and the expected frequencies.

---

1. Open a new Excel worksheet. Enter the observed frequencies in rows 1 and 2. Enter the expected frequencies in rows 4 and 5. Click in cell **A7** where the output will be placed.

	A	B	C	D	E
1	600	288	204	24	84
2	410	340	180	20	50
3					
4	550.91	342.55	209.45	24	73.09
5	459.09	285.45	174.55	20	60.91
6					
7					

2. At the top of the screen, click **Formulas** and select **Insert Function**.

3. Select the **Statistical** category. Select the **CHITEST** function. Click **OK**.

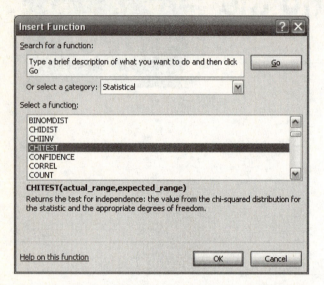

4. Complete the CHITEST dialog box as shown below. Click **OK**.

5. The function returns 1.42231E-06. This is the one-tailed probability associated with the obtained chi-square test value. To find obtained chi-square, you will be using the CHIINV function. Click in cell **A8** where the output will be placed.

7	1.42231E-06
8	

6. At the top of the screen, click **Formulas** and select **Insert Function**.

7. Select the **Statistical** category. Select the **CHIINV** function. Click **OK**.

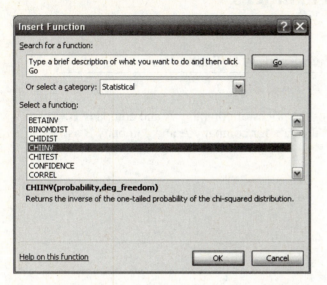

8. Complete the CHIINV dialog box as shown below. The function will accept either the numerical value of the probability or the worksheet location of the probability. For this example, you are entering the worksheet location of the probability. Click **OK**.

The function returns a chi-square test value of 32.630.

7	1.42231E-06
8	32.62958066

Section 10.3 Comparing Two Variances

► Exercise 21 (pg. 572)	Performing a Two-Sample F-Test

1. The problem asks you to compare the variances of the prices of company A and company B. Begin by opening a new Excel worksheet. Enter the data for company A and company B as shown below.

	A	B
1	Co. A	Co. B
2	250	450
3	650	250
4	285	550
5	350	400
6	550	350
7		350
8		190

2. At the top of the screen, click **Data** and select **Data Analysis** in the Analysis group. Select **F-Test Two-Sample for Variances**. Click **OK**.

If Data Analysis does not appear as a choice in the Data ribbon, you will need to load the Microsoft Excel Analysis ToolPak add-in. Follow the procedure in Section GS 8.1 before continuing.

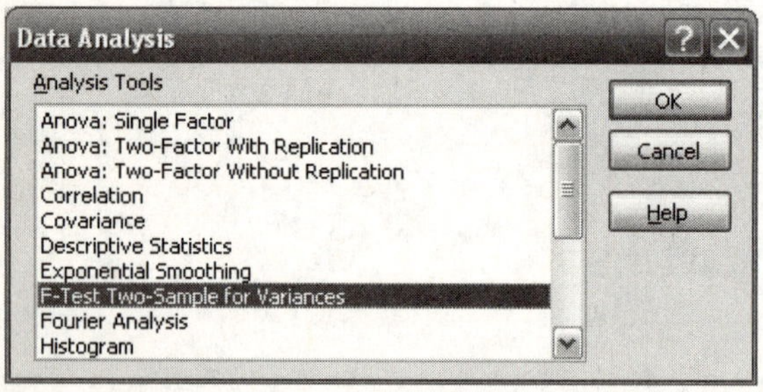

Data Analysis

Analysis Tools

Anova: Single Factor
Anova: Two-Factor With Replication
Anova: Two-Factor Without Replication
Correlation
Covariance
Descriptive Statistics
Exponential Smoothing
F-Test Two-Sample for Variances
Fourier Analysis
Histogram

OK
Cancel
Help

3. Complete the dialog box as shown at the top of the next page. Be sure to select **Labels**. Click **OK**.

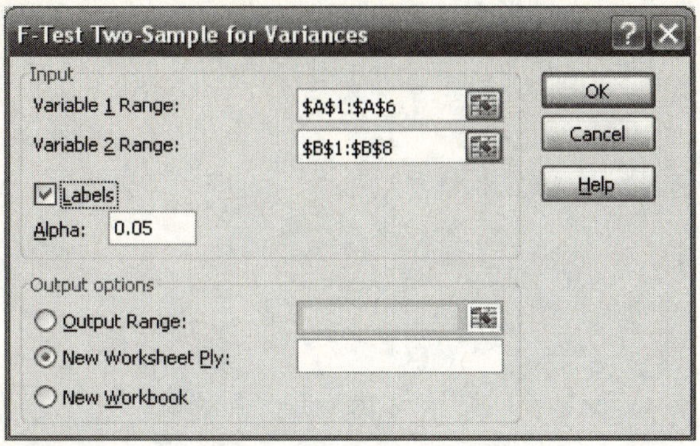

F-Test Two-Sample for Variances

Input
Variable 1 Range: $A$1:$A$6
Variable 2 Range: $B$1:$B$8
☑ Labels
Alpha: 0.05

Output options
○ Output Range:
◉ New Worksheet Ply:
○ New Workbook

[OK] [Cancel] [Help]

Your output should appear similar to the output shown below. The variance of company A prices is 30,445. The variance of company B prices is 14,490.48. Obtained F is equal to 2.101. The one-tailed probability of obtained F is 0.1988. Critical F is equal to 4.534.

	A	B	C
1	F-Test Two-Sample for Variances		
2			
3		Co. A	Co. B
4	Mean	417	362.8571
5	Variance	30445	14490.48
6	Observations	5	7
7	df	4	6
8	F	2.101035	
9	P(F<=f) one-tail	0.198784	
10	F Critical one-tail	4.533677	

◄

Section 10.4 Analysis of Variance

► Example 2 (pg. 579)   Using Technology to Perform ANOVA Tests

1. Open worksheet "Earnings" in the Chapter 10 folder.

2. At the top of the screen, click **Data** and select **Data Analysis** in the Analysis group. Select **ANOVA: Single Factor** and click **OK**.

If Data Analysis does not appear as a choice in the Data ribbon, you will need to load the Microsoft Excel Analysis ToolPak add-in. Follow the procedure in Section GS 8.1 before continuing.

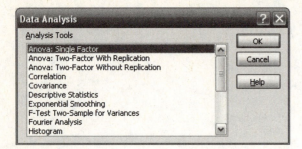

3. Complete the ANOVA: Single Factor dialog box as shown below. Be sure to select **Labels in first row** and change **Alpha** to **0.10**. Click **OK**.

Make column A wider so that you can read all the labels. Your output should appear similar to the output displayed below.

	A	B	C	D	E	F	G
1	Anova: Single Factor						
2							
3	SUMMARY						
4	Groups	Count	Sum	Average	Variance		
5	Actor	15	375	25	283.5714		
6	Athlete	10	380	38	778.6667		
7	Musician	11	541	49.18182	938.5636		
8							
9							
10	ANOVA						
11	Source of Variation	SS	df	MS	F	P-value	F crit
12	Between Groups	3766.364	2	1883.182	3.051763	0.060802	2.47099
13	Within Groups	20363.64	33	617.0799			
14							
15	Total	24130	35				

▶ Exercise 5 (pg. 581)	Testing the Claim That the Mean Toothpaste Costs Per Ounce Are Different

1. Open worksheet "Ex10_4-5" in the Chapter 10 folder.

2. At the top of the screen, click **Data** and select **Data Analysis** in the Analysis group. Select **ANOVA: Single Factor** and click **OK**.

If Data Analysis does not appear as a choice in the Data ribbon, you will need to load the Microsoft Excel Analysis ToolPak add-in. Follow the procedure in Section GS 8.1 before continuing.

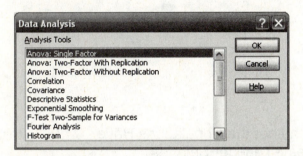

3. Complete the ANOVA: Single Factor dialog box as shown below. Click **OK**.

Make column A wider so that you can read all the labels. Your output should appear similar to the output displayed at the top of the next page.

	A	B	C	D	E	F	G
1	Anova: Single Factor						
2							
3	SUMMARY						
4	*Groups*	*Count*	*Sum*	*Average*	*Variance*		
5	Very Good	12	6.43	0.535833	0.103827		
6	Good	12	9.92	0.826667	0.443152		
7	Fair	5	3.15	0.63	0.1531		
8							
9							
10	ANOVA						
11	*Source of Variation*	*SS*	*df*	*MS*	*F*	*P-value*	*F crit*
12	Between Groups	0.518373	2	0.259186	1.016546	0.375777	3.369016
13	Within Groups	6.629158	26	0.254968			
14							
15	Total	7.147531	28				

◄

Technology

► Exercise 3 (pg. 595)	Determining Whether the Populations Have Equal Variances—Teacher Salaries

1. Open worksheet "Tech10-a" in the Chapter 10 folder. For this exercise, instructions are provided only for testing the equality of the California and Ohio variances.

2. At the top of the screen, click **Data** and select **Data Analysis** in the Analysis group. Select **F-Test Two-Sample for Variances**. Click **OK**.

If Data Analysis does not appear as a choice in the Data ribbon, you will need to load the Microsoft Excel Analysis ToolPak add-in. Follow the procedure in Section GS 8.1 before continuing.

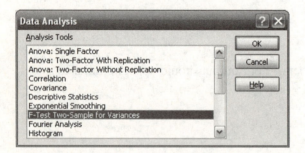

3. Complete the dialog box as shown at the top of the next page. Be sure to select **Labels**. Click **OK**.

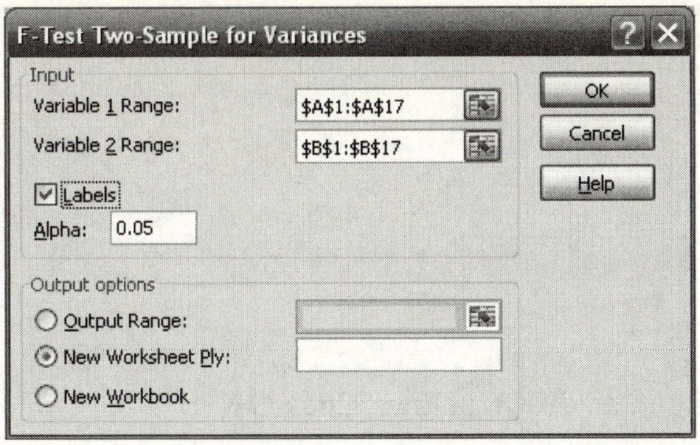

You will want to make column A wider so that you can read all the labels. Your output should appear similar to the output displayed below.

	A	B	C
1	F-Test Two-Sample for Variances		
2			
3		California	Ohio
4	Mean	63604.25	53538.5
5	Variance	58751772	61587681
6	Observations	16	16
7	df	15	15
8	F	0.953953	
9	P(F<=f) one-tail	0.464233	
10	F Critical one-tail	0.416069	

◄

► Exercise 4 (pg. 595)	**Testing the Claim That Teachers from the Three States Have the Same Mean Salary**

1. Open worksheet "Tech10-a" in the Chapter 10 folder.

   *If you have just completed Exercise 3 on page 595 and have not yet closed the Excel worksheet, return to the sheet with the data by clicking on the **Sheet1** tab near the bottom of the screen.*

2. At the top of the screen, click **Data** and select **Data Analysis** in the Analysis group. Select **ANOVA: Single Factor** and click **OK**.

   If Data Analysis does not appear as a choice in the Data ribbon, you will need to load the Microsoft Excel Analysis ToolPak add-in. Follow the procedure in Section GS 8.1 before continuing.

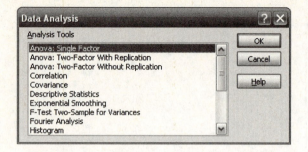

3. Complete the ANOVA: Single Factor dialog box as shown below. Click **OK**.

Make column A wider so that you can read all the labels. Your output should appear similar to the output displayed below.

	A	B	C	D	E	F	G
1	Anova: Single Factor						
2							
3	SUMMARY						
4	Groups	Count	Sum	Average	Variance		
5	California	16	1017668	63604.25	58751772		
6	Ohio	16	856616	53538.5	61587681		
7	Texas	16	726272	45392	27668644		
8							
9							
10	ANOVA						
11	Source of Variation	SS	df	MS	F	P-value	F crit
12	Between Groups	2.66E+09	2	1.33E+09	26.99154	1.98E-08	3.204317
13	Within Groups	2.22E+09	45	49336032			
14							
15	Total	4.88E+09	47				

Nonparametric Tests

CHAPTER
11

Section 11.1 The Sign Test

▶ Example 1 (pg. 600)	Using the Sign Test

If the DDXL add-in has not been loaded, you will need to load it before continuing. Follow the instructions in Section GS 8.2.

1. Open a new Excel worksheet and enter the visitors data as shown below.

	A
1	Visitors
2	1469
3	1463
4	1487
5	1579
6	1462
7	1476
8	1523
9	1620
10	1634
11	1570
12	1525
13	1568
14	1602
15	1544
16	1548
17	1492
18	1500
19	1452
20	1511
21	1649

2. Click and drag over the data range **A1:A21** so that it is highlighted.

3.   At the top of the screen, click **Add-Ins** and select **DDXL**.

4.   Select **Nonparametric Tests**. Under Function type, select **Sign Test**. Click **OK**.

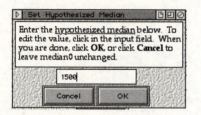

5.   Select **Visitors** in the Names and Columns Window. Then click the arrow under Quantitative Variable to select Visitors for the analysis. Click **OK**.

6.   Click **Set Hypothesized Median**. Enter a value of **1500**. Click **OK**.

7.   Select the **0.05** significance level. Select a **Right Tailed** test. Click **Compute**.

The output is displayed below. There are 12 positive differences and 7 negative differences. The *p*-value is equal to 0.1796.

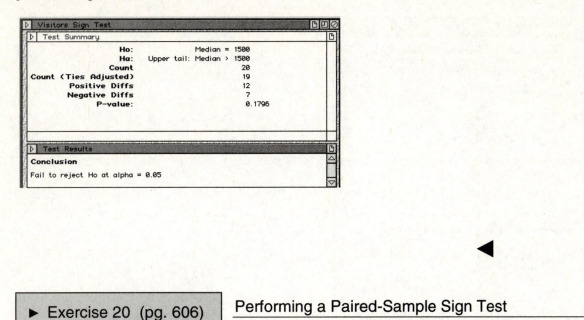

▶ **Exercise 20 (pg. 606)**   Performing a Paired-Sample Sign Test

If the DDXL add-in has not been loaded, you will need to load it before continuing. Follow the instructions in Section GS 8.2.

1. Open worksheet "Ex11_1-20" in the Chapter 11 folder.

2. Click and drag over the data range **A1:B13** so that it is highlighted.

3. At the top of the screen, click **Add-Ins** and select **DDXL**. Select **Nonparametric Tests**. Under Function type, select **Paired Sign Test**.

4. Select **Before** for the 1st Quantitative Variable. Select **After** for the 2nd Quantitative Variable. Click **OK**.

5. Select **0.05** for the significance level. Select a **Right Tailed** test. Click **Compute**.

The output is displayed below. There are 8 positive differences and 4 negative differences. The *p*-value is equal to 0.19.

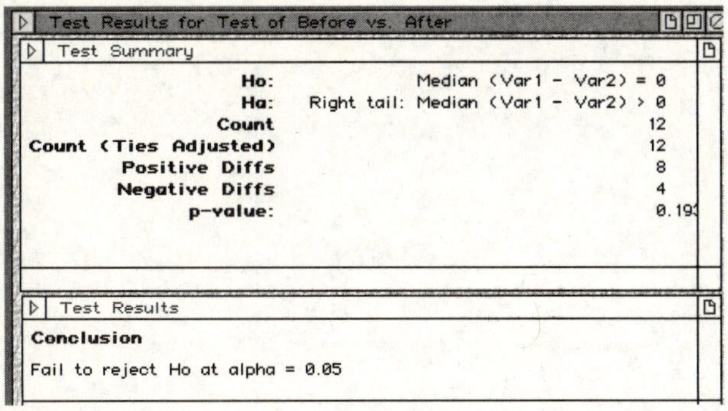

```
┌─────────────────────────────────────────────────────────────┐
│ ▷ │ Test Results for Test of Before vs. After    🗎 🗗 🗙      │
│ ┌───────────────────────────────────────────────────────┬─┐ │
│ │ ▷ │ Test Summary                                      │🗎│ │
│ │                                                         │ │
│ │              Ho:        Median (Var1 - Var2) = 0        │ │
│ │              Ha:  Right tail: Median (Var1 - Var2) > 0  │ │
│ │            Count                           12           │ │
│ │   Count (Ties Adjusted)                    12           │ │
│ │        Positive Diffs                       8           │ │
│ │        Negative Diffs                       4           │ │
│ │             p-value:                      0.19          │ │
│ │                                                         │ │
│ │                                                         │ │
│ ├───────────────────────────────────────────────────────┼─┤ │
│ │ ▷ │ Test Results                                      │🗎│ │
│ │ Conclusion                                              │ │
│ │ Fail to reject Ho at alpha = 0.05                       │ │
│ └───────────────────────────────────────────────────────┴─┘ │
└─────────────────────────────────────────────────────────────┘
```

◀

Section 11.2 The Wilcoxon Tests

▶ **Example 2 (pg. 613)**   Performing a Wilcoxon Rank Sum Test

If the DDXL add-in has not been loaded, you will need to load it before continuing. Follow the instructions in Section GS 8.2.

1. Open worksheet "Earnings" in the Chapter 11 folder. Click and drag over the data range **A1:B13** so that it is highlighted.

2. At the top of the screen, click **Add-Ins** and select **DDXL**. Select **Nonparametric Tests**. Under Function type, select **Mann Whitney Rank Sum**.

3.   Select **Male** for the 1st Quantitative Variable. Select **Female** for the 2nd Quantitative Variable. Click **OK**.

4.   Select **0.10** for the significance level. Select a **Two Tailed** test. Click **Compute**.

Your output should look similar to the output displayed below.

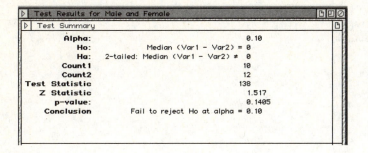

Section 11.3 The Kruskal-Wallis Test

► Example 1 (pg. 621)	Performing a Kruskal-Wallis Test

If the DDXL add-in has not been loaded, you will need to load it before continuing. Follow the instructions in Section GS 8.2.

1. Open worksheet "Crimes" in the Chapter 11 folder. The data need to be rearranged so that precinct number is in one column and number of crimes in another. The rearranged data are placed in columns D and E of the worksheet shown below. Continue in the same way until you have the rearranged the data for all three precincts.

	A	B	C	D	E
1	101st Precinct	106th Precinct	113th Precinct	Precinct	Crimes
2	60	65	69	101	60
3	52	55	51	101	52
4	49	64	70	101	49
5	52	66	61	101	52
6	50	53	67	101	50
7	48	58	65	101	48
8	57	50	62	101	57
9	45	54	59	101	45
10	44	70	60	101	44
11	56	62	63	101	56
12				106	65

2. Click and drag over the data range **D1:E31** so that it is highlighted. At the top of the screen, click **Add-Ins** and select **DDXL**. Select **Nonparametric Tests**. Under Function type, select **Kruskal Wallis**.

3. Select **Crimes** for the Response Variable. Select **Precinct** for the Factor Variable. Click **OK**.

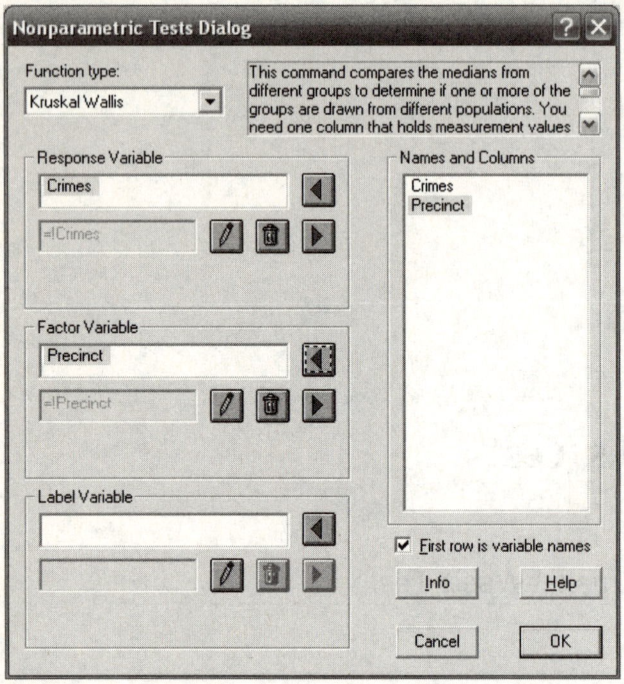

```
┌─ Nonparametric Tests Dialog ──────────────────── ? X ─┐
│                                                        │
│  Function type:           This command compares the    │
│  ┌──────────────┐ ▼       medians from different       │
│  │ Kruskal Wallis│         groups to determine if one  │
│  └──────────────┘         or more of the groups are    │
│                           drawn from different         │
│                           populations. You need one    │
│                           column that holds            │
│                           measurement values  ▼        │
│  ┌─ Response Variable ─┐  ┌─ Names and Columns ─┐      │
│  │ Crimes          │ ◄ │  │ Crimes              │      │
│  │ =!Crimes   [/][盒][►]│  │ Precinct            │      │
│  └────────────────────┘  │                     │      │
│                          │                     │      │
│  ┌─ Factor Variable ───┐ │                     │      │
│  │ Precinct        │ ◄ │ │                     │      │
│  │ =!Precinct [/][盒][►]│ │                     │      │
│  └────────────────────┘  │                     │      │
│                          │                     │      │
│  ┌─ Label Variable ────┐ │                     │      │
│  │                 │ ◄ │ │                     │      │
│  │           [/][盒][►] │ └─────────────────────┘      │
│  └────────────────────┘  ☑ First row is variable names │
│                            [Info]      [Help]          │
│                            [Cancel]    [ OK ]          │
└────────────────────────────────────────────────────────┘
```

The test results and summary are displayed below. The output also includes boxplots.

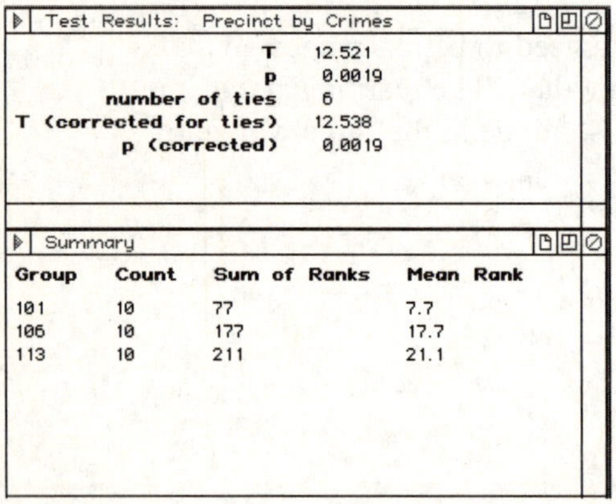

```
┌ ▷  Test Results:  Precinct by Crimes       [▣][▣]|⊘ │
│                        T    12.521                   │
│                        p     0.0019                  │
│            number of ties    6                       │
│      T (corrected for ties)  12.538                  │
│            p (corrected)     0.0019                  │
│                                                      │
└──────────────────────────────────────────────────────┘

┌ ▷  Summary                                  [▣][▣]|⊘ │
│                                                      │
│  Group   Count   Sum of Ranks   Mean Rank           │
│                                                      │
│  101      10      77             7.7                 │
│  106      10      177            17.7                │
│  113      10      211            21.1                │
│                                                      │
│                                                      │
│                                                      │
└──────────────────────────────────────────────────────┘
```

◄

Section 11.4 Rank Correlation

▶ Exercise 6 (pg. 628)	Testing a Claim about the Correlation Between Overall Score and Price

If the DDXL add-in has not been loaded, you will need to load it before continuing. Follow the instructions in Section GS 8.2.

1. Open worksheet "Ex11_4-6" in the Chapter 11 folder. Click and drag over the data range **A1:B12** so that it is highlighted.

2. At the top of the screen, click **Add-Ins** and select **DDXL**. Select **Nonparametric Tests**. Under Function type, select **Spearman Rank Test**.

3. Select **Overall score** for the x-Axis Quantitative Variable. Select **Price** for the y-Axis Quantitative Variable. Click **OK**.

4. Select **Perform a Two Tailed Test**.

The test results are shown at the bottom of the output. The t statistic is equal to 3.189. The p-value is 0.011.

```
 ▷ │ Test Results                    🗎🗗⊘
   │            Ha    Two Tailed         △
   │ t statistic            3.189
   │    p-value             0.011
   │                                    ▽
 ◁│                                  ▷◇│◇
```

◀

Technology

▶ Exercise 2 (pg. 647)	Performing a Sign Test

If the DDXL add-in has not been loaded, you will need to load it before continuing. Follow the instructions in Section GS 8.2.

1. Open worksheet "Tech11_a" in the Chapter 11 folder. Click and drag over the Midwest data, **B1:B13**, so that it is highlighted.

2. At the top of the screen, click **Add-Ins** and select **DDXL**. Select **Nonparametric Tests**. Under Function type, select **Sign Test**.

3. Select **Midwest** for the Quantitative Variable. Click **OK**.

```
 ┌──────────────────────────────────────────────────┐
 │ Nonparametric Tests Dialog                ? X     │
 ├──────────────────────────────────────────────────┤
 │ Function type:        This command computes a sign test for a  ▲│
 │                       hypothesized population median. You need a │
 │ Sign Test      ▼      column that holds quantitative values.    ▼│
 │                                                    │
 │ ┌Quantitative Variable─┐   ┌Names and Columns─┐    │
 │ │ Midwest          │ ◀▮ │   │ Midwest          │    │
 │ └──────────────────────┘   └──────────────────┘    │
 └──────────────────────────────────────────────────┘
```

4. Click **Set Hypothesized Median** and enter 30000. Click **OK**.

```
 ▷ │ Set Hypothesized Median        🗎🗗⊘
   │ Enter the hypothesized median below. To
   │ edit the value, click in the input field. When
   │ you are done, click OK, or click Cancel to
   │ leave median0 unchanged.
   │
   │        ┌─────────────┐
   │        │ 30000       │
   │        └─────────────┘
   │     ┌────────┐  ┌────────┐
   │     │ Cancel │  │   OK   │
   │     └────────┘  └────────┘
```

5.   Select **0.05** for the significance level. Select a **Right Tailed** test. Click **Compute**.

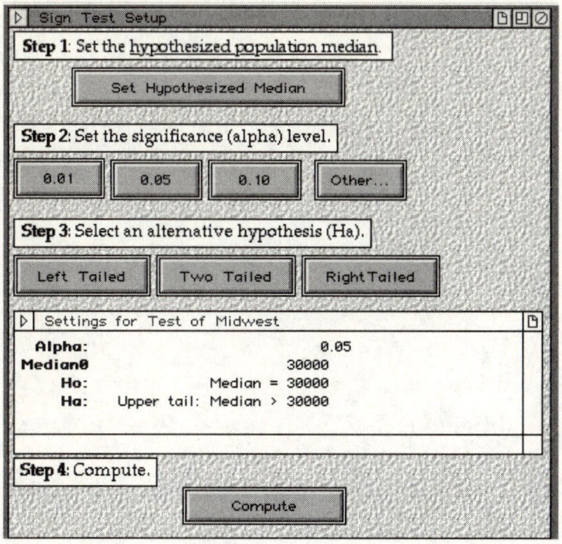

```
┌─────────────────────────────────────────────────┐
│ ▷ Sign Test Setup                        [⬚][⬛][∅]│
│ ┌─────────────────────────────────────────────┐ │
│ │ Step 1: Set the hypothesized population median.│ │
│ │       ┌───────────────────────────────┐     │ │
│ │       │   Set Hypothesized Median     │     │ │
│ │       └───────────────────────────────┘     │ │
│ │ Step 2: Set the significance (alpha) level.  │ │
│ │ ┌──────┐ ┌──────┐ ┌──────┐  ┌────────┐       │ │
│ │ │ 0.01 │ │ 0.05 │ │ 0.10 │  │ Other… │       │ │
│ │ └──────┘ └──────┘ └──────┘  └────────┘       │ │
│ │ Step 3: Select an alternative hypothesis (Ha).│ │
│ │ ┌────────────┐ ┌────────────┐ ┌────────────┐ │ │
│ │ │ Left Tailed│ │ Two Tailed │ │Right Tailed│ │ │
│ │ └────────────┘ └────────────┘ └────────────┘ │ │
│ │ ┌─────────────────────────────────────────┐  │ │
│ │ │▷ Settings for Test of Midwest         [⬚]│  │ │
│ │ │  Alpha:              0.05                │  │ │
│ │ │  Median0         30000                   │  │ │
│ │ │    Ho:        Median = 30000             │  │ │
│ │ │    Ha:   Upper tail: Median > 30000      │  │ │
│ │ └─────────────────────────────────────────┘  │ │
│ │ Step 4: Compute.                              │ │
│ │       ┌───────────────────────────────┐     │ │
│ │       │          Compute              │     │ │
│ │       └───────────────────────────────┘     │ │
│ └─────────────────────────────────────────────┘ │
└─────────────────────────────────────────────────┘
```

Your output should look similar to the output shown below.

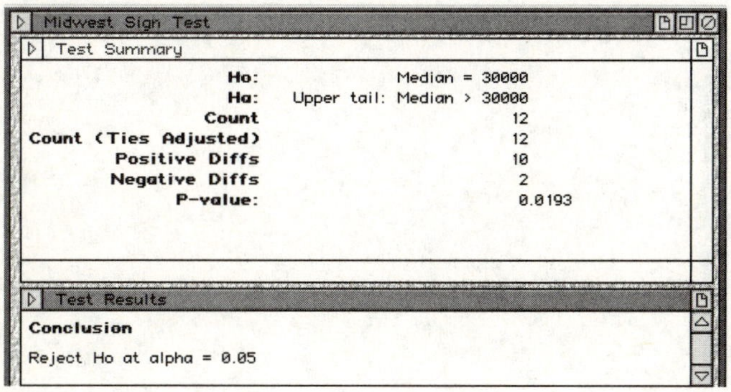

```
┌─────────────────────────────────────────────────┐
│ ▷ Midwest Sign Test                      [⬚][⬛][∅]│
│ ┌─────────────────────────────────────────────┐ │
│ │ ▷ Test Summary                            [⬚]│ │
│ │               Ho:        Median = 30000      │ │
│ │               Ha:  Upper tail: Median > 30000│ │
│ │             Count              12            │ │
│ │ Count (Ties Adjusted)          12            │ │
│ │     Positive Diffs             10            │ │
│ │     Negative Diffs              2            │ │
│ │          P-value:           0.0193           │ │
│ └─────────────────────────────────────────────┘ │
│ ┌─────────────────────────────────────────────┐ │
│ │ ▷ Test Results                            [⬚]│ │
│ │ Conclusion                                   │ │
│ │ Reject Ho at alpha = 0.05                    │ │
│ └─────────────────────────────────────────────┘ │
└─────────────────────────────────────────────────┘
```

◄

▶ **Exercise 3 (pg. 647)**   Performing a Wilcoxon Rank Sum Test

---

If the DDXL add-in has not been loaded, you will need to load it before continuing. Follow the instructions in Section GS 8.2.

---

1.   Open worksheet "Tech11_a" in the Chapter 11 folder.

2. Copy the Northeast and South data so that they are located in adjacent columns of the worksheet. For this exercise, use columns F and G.

F	G
Northeast	South
47,000	24,030
35,145	37,943
31,497	36,280
27,500	38,738
28,500	22,275
35,400	27,975
33,810	28,275
32,500	35,073
29,950	39,730
25,100	36,775
42,700	25,675
49,950	29,875

3. Click and drag over the data range **F1:G13** so that it is highlighted.

4. At the top of the screen, click **Add-Ins** and select **DDXL**. Select **Nonparametric Tests**. Under Function type, select **Mann Whitney Rank Sum**.

5. Select **Northeast** for the 1st Quantitative Variable. Select **South** for the 2nd Quantitative Variable. Click **OK**.

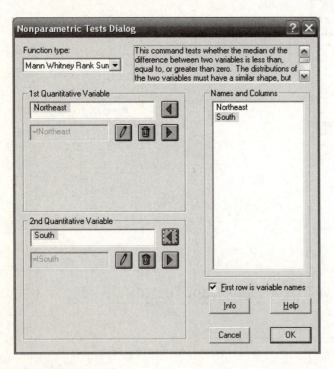

6. Select **0.05** for the significance level. Select a **Two Tailed** test. Click **Compute**.

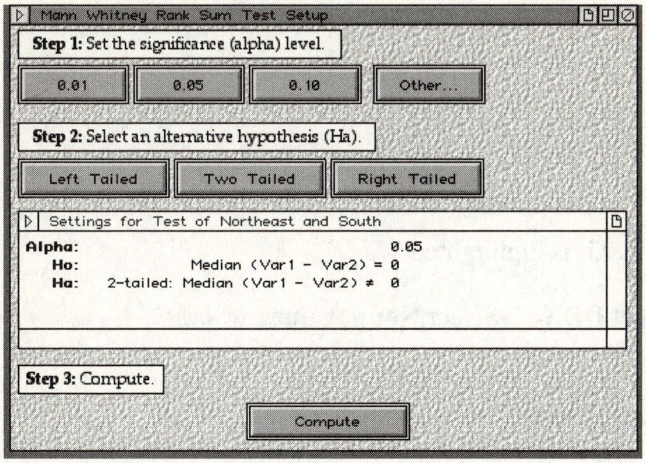

```
┌─────────────────────────────────────────────────────────────┐
│ ▷  Mann Whitney Rank Sum Test Setup                  ▣ ▣ ∅   │
│  ┌──────────────────────────────────────────────────┐       │
│  │ Step 1: Set the significance (alpha) level.       │       │
│  │  ┌────────┐  ┌────────┐  ┌────────┐  ┌─────────┐  │       │
│  │  │  0.01  │  │  0.05  │  │  0.10  │  │ Other...│  │       │
│  │  └────────┘  └────────┘  └────────┘  └─────────┘  │       │
│  │ Step 2: Select an alternative hypothesis (Ha).    │       │
│  │ ┌────────────┐ ┌────────────┐ ┌─────────────┐     │       │
│  │ │ Left Tailed│ │ Two Tailed │ │ Right Tailed│     │       │
│  │ └────────────┘ └────────────┘ └─────────────┘     │       │
│  │ ┌──────────────────────────────────────────────┐ │       │
│  │ │▷ Settings for Test of Northeast and South   ▯│ │       │
│  │ │ Alpha:                       0.05            │ │       │
│  │ │   Ho:        Median (Var1 - Var2) = 0        │ │       │
│  │ │   Ha:   2-tailed: Median (Var1 - Var2) ≠ 0   │ │       │
│  │ │                                              │ │       │
│  │ └──────────────────────────────────────────────┘ │       │
│  │ ┌──────────┐                                      │       │
│  │ │ Step 3: │ Compute.                              │       │
│  │ └──────────┘     ┌──────────────┐                 │       │
│  │                  │   Compute    │                 │       │
│  │                  └──────────────┘                 │       │
│  └──────────────────────────────────────────────────┘       │
└─────────────────────────────────────────────────────────────┘
```

Your output should look similar to the output shown below.

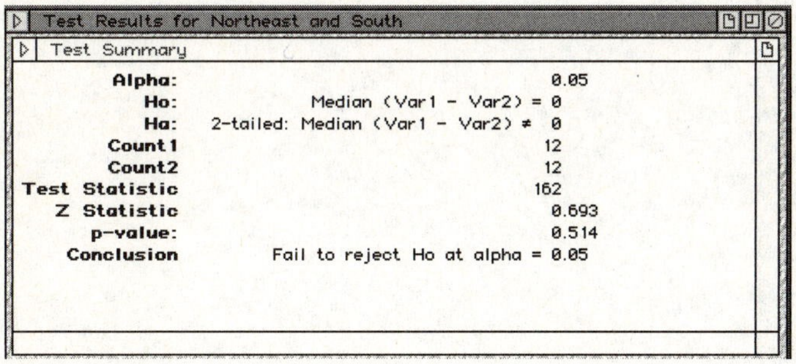

```
┌─────────────────────────────────────────────────────────────┐
│ ▷  Test Results for Northeast and South              ▣ ▣ ∅   │
│ ┌───────────────────────────────────────────────────────┐   │
│ │▷ Test Summary                                        ▯│   │
│ │        Alpha:                        0.05             │   │
│ │          Ho:        Median (Var1 - Var2) = 0          │   │
│ │          Ha:   2-tailed: Median (Var1 - Var2) ≠ 0     │   │
│ │       Count 1                         12              │   │
│ │       Count 2                         12              │   │
│ │ Test Statistic                        162             │   │
│ │    Z Statistic                       0.693            │   │
│ │       p-value:                       0.514            │   │
│ │    Conclusion      Fail to reject Ho at alpha = 0.05  │   │
│ │                                                       │   │
│ └───────────────────────────────────────────────────────┘   │
└─────────────────────────────────────────────────────────────┘
```

◀

▶ **Exercise 4 (pg. 647)**   Performing a Kruskal-Wallis Test

---

If the DDXL add-in has not been loaded, you will need to load it before continuing. Follow the instructions in Section GS 8.2.

---

1. Open worksheet "Tech11_a" in the Chapter 11 folder.

---

*If you have just completed Exercise 2 or 3 on page 647 and have not yet closed the worksheet, click on the **Sheet 1** tab at the bottom of the screen to return to the worksheet containing the data.*

---

2.   The data need to be rearranged for this analysis. See the instructions in this chapter for Example 1 in Section 11.3. For this exercise, the rearranged data were placed in columns F and G. The first few rows are shown below.

F	G
Region	Income
Northeast	47,000
Northeast	35,145
Northeast	31,497

3.   Click and drag over the data range **F1:G49** so that it is highlighted.

4.   At the top of the screen, click **Add-Ins** and select **DDXL**. Select **Nonparametric Tests**. Under Function type, select **Kruskal Wallis**.

5.   Select **Income** for the Response Variable. Select **Region** for the Factor Variable. Click **OK**.

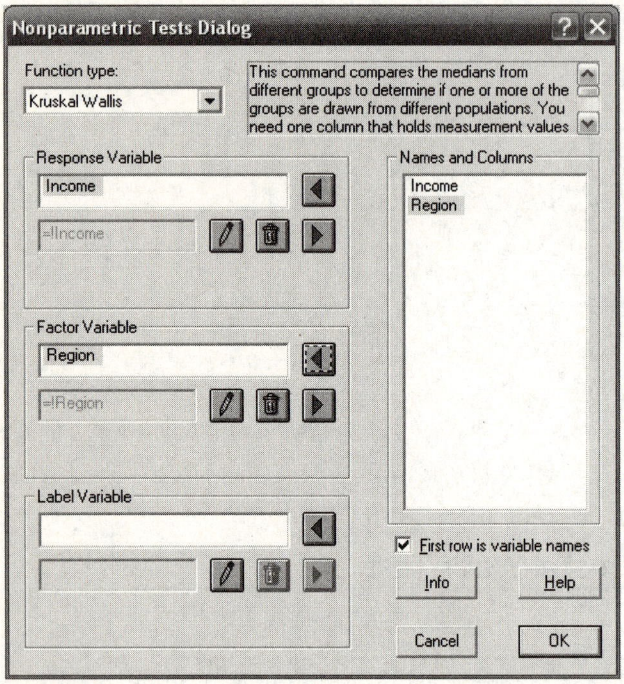

The test results, summary, and boxplots are displayed below.

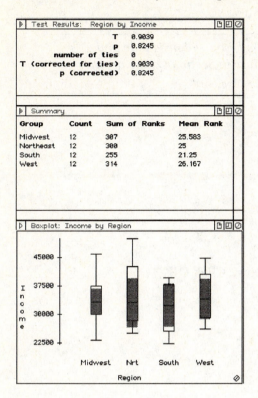

▶ **Exercise 5 (pg. 647)**   Performing a One-Way ANOVA

1.   Open worksheet "Tech11_a" in the Chapter 11 folder.

*If you have just completed Exercise 2, 3, or 4 on page 647 and have not yet closed the worksheet, click on the **Sheet1** tab at the bottom of the screen to return to the worksheet containing the data.*

2.   At the top of the screen, click **Data** and select **Data Analysis**. Select **Anova: Single Factor** and click **OK**.

If Data Analysis does not appear as a choice in the Data ribbon, you will need to load the Microsoft Excel Analysis ToolPak add-in. Follow the procedure in Section GS 8.1 before continuing.

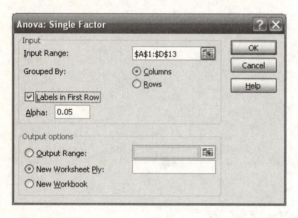

3.   Complete the Anova: Single Factor dialog box as shown below. Click **OK**.

Adjust the column width so that you can read all the labels. Your output should appear similar to the output shown below.

	A	B	C	D	E	F	G
1	Anova: Single Factor						
2							
3	SUMMARY						
4	Groups	Count	Sum	Average	Variance		
5	Northeast	12	419052	34921	60804386		
6	Midwest	12	411888	34324	39711468		
7	South	12	382644	31887	38506904		
8	West	12	415308	34609	35204675		
9							
10							
11	ANOVA						
12	Source of Variation	SS	df	MS	F	P-value	F crit
13	Between Groups	69265161	3	23088387	0.530075	0.664003	2.816466
14	Within Groups	1.92E+09	44	43556858			
15							
16	Total	1.99E+09	47				